ZIML Math Competition Book

Division E 2017-2018

Areteem Institute

ZIML Math Competition Book Division E 2017-18

Edited by   John Lensmire
            David Reynoso
            Kevin Wang
            Kelly Ren

Copyright © 2018 ARETEEM INSTITUTE

WWW.ARETEEM.ORG

PUBLISHED BY ARETEEM PRESS

ISBN: 1-944863-27-3
ISBN-13: 978-1-944863-27-2

First printing, October 2018.

TITLES PUBLISHED BY ARETEEM PRESS

Cracking the High School Math Competitions (and Solutions Manual) - Covering AMC 10 & 12, ARML, and ZIML
Mathematical Wisdom in Everyday Life (and Solutions Manual) - From Common Core to Math Competitions
Geometry Problem Solving for Middle School (and Solutions Manual) - From Common Core to Math Competitions
Fun Math Problem Solving For Elementary School (and Solutions Manual)

ZIML MATH COMPETITION BOOK SERIES

ZIML Math Competition Book Division E 2016-2017
ZIML Math Competition Book Division M 2016-2017
ZIML Math Competition Book Division H 2016-2017
ZIML Math Competition Book Jr Varsity 2016-2017
ZIML Math Competition Book Varsity Division 2016-2017
ZIML Math Competition Book Division E 2017-2018
ZIML Math Competition Book Division M 2017-2018
ZIML Math Competition Book Division H 2017-2018
ZIML Math Competition Book Jr Varsity 2017-2018
ZIML Math Competition Book Varsity Division 2017-2018

MATH CHALLENGE CURRICULUM TEXTBOOKS SERIES

Math Challenge I-A Pre-Algebra and Word Problems
Math Challenge I-B Pre-Algebra and Word Problems
Math Challenge I-C Algebra
Math Challenge II-A Algebra
Math Challenge II-B Algebra
Math Challenge III Algebra
Math Challenge I-A Geometry
Math Challenge I-B Geometry
Math Challenge I-C Topics in Algebra
Math Challenge II-A Geometry
Math Challenge II-B Geometry

Math Challenge III Geometry
Math Challenge I-B Counting and Probability
Math Challenge II-A Combinatorics
Math Challenge I-B Number Theory
Math Challenge II-A Number Theory

COMING SOON FROM ARETEEM PRESS

Fun Math Problem Solving For Elementary School Vol. 2 (and Solutions Manual)
Counting & Probability for Middle School (and Solutions Manual)
- From Common Core to Math Competitions
Number Theory Problem Solving for Middle School (and Solutions Manual) - From Common Core to Math Competitions
Other volumes in the **Math Challenge Curriculum Textbooks Series**

The books are available in paperback and eBook formats (including Kindle and other formats).

To order the books, visit https://areteem.org/bookstore.

Contents

Introduction

Each month during the school year, Areteem Institute hosts the online Zoom International Math League (ZIML) competitions. Students can compete in one of five divisions based on their age and mathematical level (details shown on Page 9).

This book contains the problems, answers, and full solutions from the nine ZIML Division E Competitions held during the 2017-2018 School Year. It is divided into three parts:

1. The complete Division E ZIML Competitions (20 questions per competition) from October 2017 to June 2018.
2. The solutions for each of the competitions, including detailed work and helpful tricks.
3. An appendix including the topics and knowledge points covered for Division E, a glossary including common mathematical terms, and answer keys for each of the competitions so students can easily check their work.

The questions found on the ZIML competitions are meant to test your problem solving skills and train you to apply the knowledge you know to many different applications. We hope you enjoy the problems!

About Zoom International Math League

The Zoom International Math League (ZIML) has a simple goal: provide a platform for students to build and share their passion for math and other STEM fields with students from around the globe. Started in 2008 as the Southern California Mathematical Olympiad, ZIML has a rich history of past participants who have advanced to top tier colleges and prestigious math competitions, including American Math Competitions, MATHCOUNTS, and the International Math Olympaid.

The ZIML Core Online Programs, most available with a free account at ziml.areteem.org, include:

- **Daily Magic Spells:** Provides a problem a day (Monday through Friday) for students to practice, with full solutions available the next day.
- **Weekly Brain Potions:** Provides one problem per week posted in the online discussion forum at ziml.areteem.org. Usually the problem does not have a simple answer, and students can join the discussion to share their thoughts regarding the scenarios described in the problem, explore the math concepts behind the problem, give solutions, and also ask further questions.
- **Monthly Contests:** The ZIML Monthly Contests are held the first weekend of each month during the school year (October through June). Students can compete in one of 5 divisions to test their knowledge and determine their strengths and weaknesses, with winners announced after the competition.
- **Math Competition Practice:** The Practice page contains sample ZIML contests and an archive of AMC-series tests for online practice. The practices simulate the real contest environment with time-limits of the contests automatically controlled by the server.
- **Online Discussion Forum:** The Online Discussion Forum

is open for any comments and questions. Other discussions, such as hard Daily Magic Spells or the Weekly Brain Potions are also posted here.

These programs encourage students to participate consistently, so they can track their progress and improvement each year.

In addition to the online programs, ZIML also hosts onsite Local Tournaments and Workshops in various locations in the United States. Each summer, there are onsite ZIML Competitions at held at Areteem Summer Programs, including the National ZIML Convention, which is a two day convention with one day of workshops and one day of competition.

ZIML Monthly Contests are organized into five divisions ranging from upper elementary school to advanced material based on high school math.

- **Varsity:** This is the top division. It covers material on the level of the last 10 questions on the AMC 12 and AIME level. This division is open to all age levels.
- **Junior Varsity:** This is the second highest competition division. It covers material at the AMC 10/12 level and State and National MathCounts level. This division is open to all age levels.
- **Division H:** This division focuses on material from a standard high school curriculum. It covers topics up to and including pre-calculus. This division will serve as excellent practice for students preparing for the math portions of the SAT or ACT. This division is open to all age levels.
- **Division M:** This division focuses on problem solving using math concepts from a standard middle school math curriculum. It covers material at the level of AMC 8 and School or Chapter MathCounts. This division is open to all students who have not started grade 9.

- **Division E:** This division focuses on advanced problem solving with mathematical concepts from upper elementary school. It covers material at a level comparable to MOEMS Division E. This division is open to all students who have not started grade 6.

This problem book features the Division E Contests. For a detailed list of topics covered for Division E see p.149 in the Appendix.

About Areteem Institute

Areteem Institute is an educational institution that develops and provides in-depth and advanced math and science programs for K-12 (Elementary School, Middle School, and High School) students and teachers. Areteem programs are accredited supplementary programs by the Western Association of Schools and Colleges (WASC). Students may attend the Areteem Institute through these options:

- Live and real-time face-to-face online classes with audio, video, interactive online whiteboard, and text chatting capabilities;
- Self-paced classes by watching the recordings of the live classes;
- Short video courses for trending math, science, technology, engineering, English, and social studies topics;
- Summer Intensive Camps on prestigious university campuses and Winter Boot Camps;
- Practice with selected daily problems for free, and monthly ZIML competitions at ziml.areteem.org.

The Areteem courses are designed and developed by educational experts and industry professionals to bring real world applications into STEM education. The programs are ideal for students who wish to build their mathematical strength in order to excel academically and eventually win in Math Competitions (AMC, AIME, USAMO, IMO, ARML, MathCounts, Math Olympiad, ZIML, and other math leagues and tournaments, etc.), Science Fairs (County Science Fairs, State Science Fairs, national programs like Intel Science and Engineering Fair, etc.) and Science Olympiad, or purely want to enrich their academic lives by taking more challenges and developing outstanding analytical, logical thinking and creative problem solving skills.

Since 2004 Areteem Institute has been teaching with methodology that is highly promoted by the new Common Core State Standards: stressing the conceptual level understanding of the math concepts, problem solving techniques, and solving problems with real world applications. With the guidance from experienced and passionate professors, students are motivated to explore concepts deeper by identifying an interesting problem, researching it, analyzing it, and using a critical thinking approach to come up with multiple solutions.

Thousands of math students who have been trained at Areteem achieved top honors and earned top awards in major national and international math competitions, including Gold Medalists in the International Math Olympiad (IMO), top winners and qualifiers at the USA Math Olympiad (USAMO/JMO), and AIME, top winners at the Zoom International Math League (ZIML), and top winners at the MathCounts National. Many Areteem Alumni have graduated from high school and gone on to enter their dream colleges such as MIT, Cal Tech, Harvard, Stanford, Yale, Princeton, U Penn, Harvey Mudd College, UC Berkeley, UCLA, etc. Those who have graduated from colleges are now playing important roles in their fields of endeavor.

Further information about Areteem Institute, as well as updates and errata of this book, can be found online at http://www.areteem.org.

Acknowledgments

This book contains the Online ZIML Division E Problems from the 2017-18 school year. These problems were created and compiled by the staff of Areteem Institute. These problems were inspired by questions from the Areteem Math Challenge Courses, past questions on the ACT/SAT/GRE, past math competitions, math textbooks, and countless other resources and people encountered by the Areteem Curriculum Department in their life devoted to math. We thank all these sources for growing and nurturing our passion for math.

The Areteem staff, including John Lensmire, David Reynoso, Kevin Wang, and Kelly Ren, are the main contributors who compiled, edited, and reviewed this book.

Lastly, thanks to all the students who have participated and continue to participate in the Zoom International Math League. Your dedication to the Daily Magic Spells and Monthly Contests makes all of this possible, and we hope you continue to enjoy ZIML for years to come!

1. ZIML Contests

This part of the book contains the Division E ZIML Contests from the 2017-18 School Year. There were nine monthly competitions, held on the dates found below:

- October 6-8
- November 3-5
- December 1-3
- January 5-7
- February 2-4
- March 2-4
- April 6-8
- May 4-6
- June 1-3

1.1 ZIML October 2017 Division E

Below are the 20 Problems from the Division E ZIML Competition held in October 2017.

The answer key is available on p.158 in the Appendix.

Full solutions to these questions are available starting on p.86.

Problem 1
If 5 mice can eat a whole block of cheese in 2 hours, how many hours would it take 1 mouse to eat a whole block of cheese?

Problem 2
Lola is 4 years older than her sister Dola. Five years ago Lola's age was twice Dola's age. How old is Lola now?

Problem 3
Brandon just bought some pencils for his drawing class. He bought 15 pencils that were either type H or type HB. Each type H pencil cost $2 and each type HB pencil cost $3. If Brandon spent $37 in total, how many type HB pencils did he buy?

Problem 4

In the following diagram, the small square in the lower-right has a side length of 4 and the big square on the left has a side length of 10.

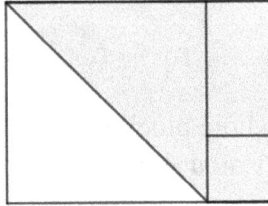

What is the area of the shaded region?

Problem 5

Drake rolls two dice. One of them was a normal 6-sided die with the numbers 1, 2, 3, 4, 5, and 6. The other die was also 6-sided but only had the numbers 4, 5 and 6 (each repeated twice). How many different sums could Drake get if he adds up the result shown on both dice?

Problem 6

Gladys baked two cakes for her daughter's birthday. As one cake was bigger, it spent a little more time in the oven than the other cake. On average, the cakes were in the oven for 45 minutes. If the larger cake needed 10 more minutes in the oven, how much time was the small cake in the oven in minutes?

Problem 7

Consider the diagram below.

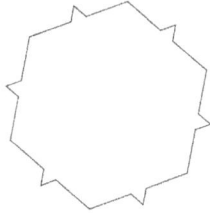

How many lines of symmetry does the figure have?

Problem 8

Consider the pattern of numbers starting with

$$5, 10, 12, 24, 26, 52, 54, \ldots$$

What is the 12th number in the sequence?

Problem 9

In the following diagram all the small rectangles are the same size.

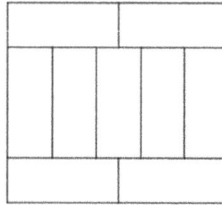

If the length of one of the rectangles is 15, what is the perimeter of the whole picture?

Problem 10

Charly got in trouble with his mom because he and his friend Mandy ate a whole bag of chocolate covered raisins one evening. They ate them so fast that they lost count of how many each of them ate. They just remember that Mandy ate 4 right after they found the bag, and, after those 4, Mandy ate three times as many as Charly. If the bag had 124 chocolate covered raisins when they found it, how many did Charly eat?

Problem 11

Dorothy went to visit her parents in Kansas. She noticed that the pictures that usually hung on the wall had fallen off. There were 7 pictures in total, including one of the whole family. She decided to help put the pictures up again but she didn't remember the order of the pictures, just that they were arranged in a line. If Dorothy wanted to make sure that the picture of the whole family was in the middle, in how many ways could she order the pictures?

Problem 12

Tyler was building some model stairs using building blocks. Each building block is a cube with length 2 cm per side. The diagram shows one of the models he built.

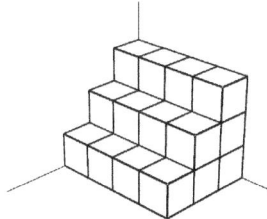

What is the volume of the whole model in cubic centimeters? Assume there are no gaps between the blocks you cannot see in the diagram.

Problem 13

Gus will participate in a race that is 2 kilometers long and wants to estimate the time it will take him to complete it. He knows his average running speed is 5 miles per hour. If he runs at that same speed, how much time (in minutes) would it take Gus to complete the 2 km run? Use 1 mi = 1.6 km and round your answer to the nearest whole minute if necessary.

Problem 14

Katya is building a kite with some colored paper, thread, and two sticks as shown in the picture. (Note the picture is not necessarily drawn to scale.)

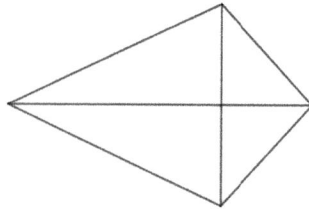

If the sticks Katya is using have length 30 cm and 20 cm, how many square centimeters of paper will she need to cover the kite?

Problem 15

What is the largest 2-digit number that leaves a remainder of 11 when divided by 13?

Problem 16

Jerry is building a box with an open top with length 30 cm, width 20 cm, and height 25 cm. He then wants to cover all the exterior and the interior of the box with colored paper. How many square centimeters of colored paper will Jerry need to cover the box?

Problem 17

Diego had 2 huge pieces of paper. He wanted to cut them into small pieces to make confetti. He started by cutting each piece into four equal smaller pieces. Then he cut each of those pieces into four pieces. Finally he cut all the pieces he had into four equal pieces one last time. How many pieces of paper did Diego have at the end?

Problem 18

Ms. McNeilly brought a bag of different colored marbles to class. She told her students that the bag had 5 green marbles, 4 red marbles, and 6 blue marbles. She then gave each student a bar of chocolate and a challenge: they had to come and grab some marbles from the bag without looking, making sure that they got at least one of the green marbles. She promised a second bar of chocolate to the students that could do this by taking as few marbles as possible. If you were Ms. McNeilly's student, how many marbles would you need to take to make sure you win the extra chocolate bar?

Problem 19

Doris found a box with 117 colored pencils in her attic and she wants to give them to her friends. She wants to be fair so she will bring exactly enough pencils so that each of her friends gets the same number of pencils and there will be no pencils remaining. She was planning on giving them to 5 of her friends, but heard that her two friends that are twins might have gone on vacation. Doris wants to be prepared regardless of whether all 5 friends or only 3 of them (not the twins) show up. If Doris wants to bring as many pencils as possible, how many should she bring?

Problem 20

An ant started on one vertex of a cube with side length 4 cm. She started walking along the edges of the cube, never walking on the same edge twice. If she walked a total of 23 cm, how many vertices of the cube did she visit (including the first)?

1.2   ZIML November 2017 Division E

Below are the 20 Problems from the Division E ZIML Competition held in November 2017.
The answer key is available on p.159 in the Appendix.
Full solutions to these questions are available starting on p.92.

Problem 1
Ty used 22 identical squared tiles to create the figure below.

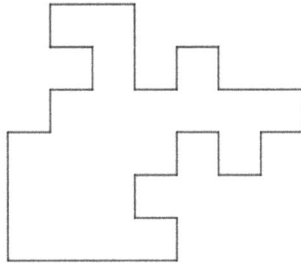

If each of the tiles has area 1 square inch, how many inches is the perimeter of the figure?

Problem 2
At first there are 12 people in a room. 4 more people come in, one by one, and shake hands with everyone that is already in the room. How many hand shakes occur in total as the 4 new people arrive?

Problem 3

Eight trees are equally spaced on one side of a straight road. The distance from the third tree to the fifth is 70 feet. How many feet are between the first and last trees?

Problem 4

Paula noticed that the faucet on her kitchen was broken and was dripping some drops of water. She counted how many drops fell to the sink each second and recorded her results in a table:

Second	# of drops
1	2
2	3
3	1
4	2
5	2
6	3
7	1
8	2
⋮	⋮

Paula noticed that the pattern seemed to repeat every 4 seconds. If this pattern continues, how many drops of water will fall to the sink after 2 minutes?

Problem 5

A rectangle has area 42. If both the length and the width of the rectangle are whole numbers, what is the largest possible perimeter for the rectangle?

Problem 6

Ala and Dylan start running from the same spot in the same direction. Ala runs at 7 mph and Dylan runs at 9 mph. If they keep running at this speed for 1.5 hours, how many miles apart will they be?

Problem 7

Arrange several equilateral triangles, all of whose side lengths are 2 cm, to form a long parallelogram, as demonstrated in the diagram.

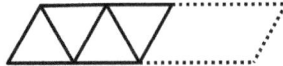

Assume the perimeter of the long parallelogram is 44 cm, how many triangles are there?

Problem 8

Sarah and her friends sit around a circular table, and start counting off numbers, clockwise, beginning with 1, and continue until 50 is counted. If there are between 10 and 20 students sitting around the table and the numbers 5 and 50 are counted by the same friend, how many friends are there in total?

Problem 9

Casey's shop class is making a golf trophy. He has to paint 300 dimples on a golf ball. If it takes him 2 seconds to paint one dimple, how many minutes will he need to do his job?

Problem 10

Sam decided it was time to be healthy. He started doing crunches every morning and every day he increased the number of crunches. The first day he did 10 crunches, the second day he adds 2 more, for a total of $10 + 2 = 12$ crunches. On the third day he added $2 \times 2 = 4$ more, so he did $12 + 4 = 16$ crunches. Each of the following days he continued to add twice as many crunches as he added the previous day to his routine. If Sam follows this pattern for 5 days, how many crunches did he do on the fifth day?

Problem 11

At Joyce's Halloween party everyone came as ghosts or were-wolves. At first there were 37 monsters at the Halloween party. Around midnight 4 ghosts left and 5 more werewolves arrived. There were then 2 more werewolves than ghosts. How many ghosts were at the party before midnight?

Problem 12

The diagram below shows three overlapped squares.

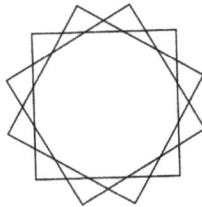

How many lines of symmetry does the figure have?

Problem 13

Luke and Cindy each have candy bars. Luke's candy bar is 2 inches wide and 10 inches long. Cindy's is 4 inches wide and 8 inches long. Cindy eats half of her candy bar. What percent of his candy bar should Luke eat so that he will eat the same amount of candy as Cindy? Round your answer to the nearest percent if necessary. For example if you get 49.3%, input an answer of 49.

Problem 14

Suppose a car travels in segments that are described in the table below:

Segment	Distance (miles)	Time (hours)
1	50	1
2	210	3
3	100	2

What is the average speed of the truck? Express your answer in miles per hour and round to the nearest tenth if necessary.

Problem 15

Jerry was building solid shapes as part of geometry class. He had already folded up 6 squares to make a cube when his teacher asked him to build a new shape. Jerry folded up the 4 triangles shown below to make a solid, called a tetrahedron.

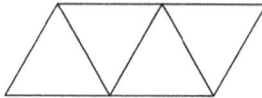

How many edges does a tetrahedron have?

Problem 16

The product of the two 7-digit numbers 3030303 and 5050505 has hundreds digit A and tens digit B. What is the sum $A + B$?

Problem 17

Melissa drove for 90 minutes at a rate of 50 miles per hour and for 30 minutes at 60 miles per hour. How many miles did she travel in total?

Problem 18

What is the largest 4-digit integer that is divisible by 3 and 5 that only uses the digits of 3 and 5?

Problem 19

Ten cubic feet hold about 75 gallons of water. Robert has a pool that is 15 feet wide, 20 feet long, and 7 feet deep. How many gallons of water does William need to fill his pool?

Problem 20

One pipe can drain a bathtub in 30 minutes. The hot water tap alone can fill in the tub in 20 minutes and the cold water tap alone can fill in the tub in 15 minutes. If both taps and the drain are open, how many minutes would it take for the bathtub to be completely full?

1.3 ZIML December 2017 Division E

Below are the 20 Problems from the Division E ZIML Competition held in December 2017.

The answer key is available on p.160 in the Appendix.

Full solutions to these questions are available starting on p.99.

Problem 1

There are 3 cows and 4 horses on the farm. Each cow eats 16 bundles of hay per day, while each horse eats 14 bundles of hay each day (the farmer makes small bundles). How many bundles of hay must the farmer keep ready each week?

Problem 2

If the tens digit of a 3-digit number is 2, the ones digit is twice the tens digit, and the hundreds digit is twice the ones digit, what is the number?

Problem 3

In the following figure the big square has side length 20 inches and the little squares have side length 10 inches.

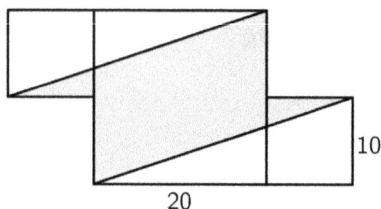

The shaded region has an area that is $N\%$ of the full figure, where N is a whole number. What is N? Round your answer to the nearest integer if necessary.

Problem 4

Anton was bored while waiting for his mom to pick him up from school so he repeatedly wrote the number 1323556787 in a piece of paper producing the following pattern

$$1323556787132355678713\ldots$$

When his mom finally came, he had written 167 digits in total. What was the last digit he wrote?

Problem 5

What is the value of the sum

$$-5 + 10 - 15 + 20 - 25 + 30 - 35 + 40 - 45 \cdots + 200?$$

Problem 6

In a far-off land 3 jars of honey can be traded for 5 fish, and 2 fish can be traded for 6 potatoes. How many potatoes are worth the same as one jar of honey?

Problem 7

In the diagram, points A, B, C, D are in the middle of the sides of the rectangle.

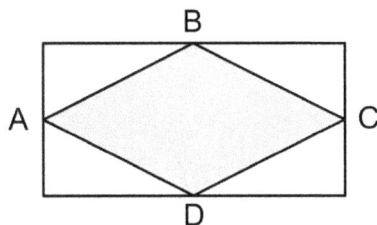

Suppose the area of the whole rectangle is 30. What is the area of the shaded region?

Problem 8

Patrick had three exams: history, math, and literature, and got a total score of 250 points. The score he got in the literature exam was 20 points less than the score he got in math, and was 10 points more than the score he got in the history exam. How many points did he receive on his math exam?

Problem 9

At Ashland Middle School 40% of the students who play volleyball are boys. If 180 students (boys and girls) play volleyball, how many girls play volleyball?

Problem 10

Adam, Bob and Cynthia each drew the shapes shown in the diagram below.

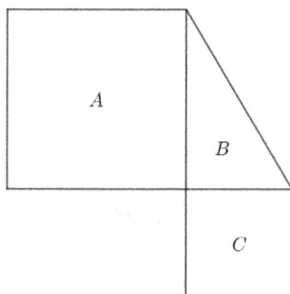

Adam and Cynthia drew squares, while Bob drew a triangle. If the area of Adam's square is 25 and the area of Bob's triangle is 7.5, what is the area of Cynthia's triangle? Round your answer to the nearest tenth if necessary.

Problem 11

You and your friends are playing Monopoly. You need to pay another player $654 and you have 16 $1 bills, 12 $5 bills, 20 $20 bills, 3 $100 bills, and 2 $500 bills. If you want to use the least number of bills possible to pay, how many bills would you need?

Problem 12

In the diagram below 8 congruent rectangles are put together to form a bigger rectangle.

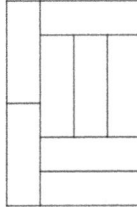

If the perimeter of each of the small rectangles is 16, what is the perimeter of the big rectangle?

Problem 13

You and your brother were playing with some marbles. At the beginning you had 103 marbles, and your brother had 92. After playing for a while you lost some marbles (so they were given to your brother) and now your brother has 4 times as many marbles as you. How many marbles did you lose to your brother?

Problem 14

Frank and Lacey were on a huge park with several ice cream shops inside it. They were trying to figure out where was the closest one. They found a sign that had the coordinates of some places of interest in the park, including the ice cream shops. The sign said they were currently standing on $(5,7)$, and listed ice cream shops on the points $(2,9)$, $(3,5)$, $(3,2)$ and $(5,2)$. What is the sum of the x-coordinate and y-coordinate of the closest ice cream shop?

Problem 15

What is the minimum number of small squares that must be colored black so that a line of symmetry lies on the diagonal *BD* of square *ABCD*?

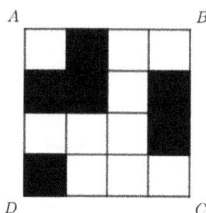

Problem 16

Greta and Hans need to gather some insects for their "Introduction to Magical Animals" class. Greta is in charge of gathering dimplies, which are flying insects that have 3 legs. Hans is in charge of gathering rumplies, which are crawling insects that have 5 legs. Their teacher wants them to dress all of the insects with hats and little shoes for a Christmas show. He provided them with exactly 24 tiny hats and 106 tiny shoes, and they must use all of them to complete their assignment successfully. If each insect has only 1 head, how many rumplies should Hans gather?

Problem 17

Ali, Bonnie, Carlo and Dianna are going to drive together to a nearby theme park. The car they are using has four seats: one driver's seat (in the front), one front passenger seat, and two back seats. Bonnie and Carlo are the only two who can drive the car. If Ali always sits next to Bonnie, how many possible seating arrangements are there?

Problem 18

In the following diagram we have A squares and B rectangles. Remember that a square is a special type of rectangle.

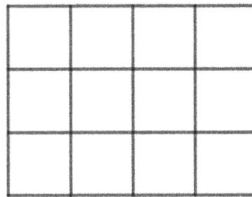

What is $B - A$?

Problem 19

You are the first in line to buy tickets for a raffle at your school. They have tickets numbered $001, 002, 003, \ldots, 200$. Your lucky number is 3, so you want to buy all the tickets that have exactly one 3. How many tickets will you buy?

Problem 20

While Rudolph waited with the other reindeer on a roof, he noticed that the Christmas lights on a house blinked following a pattern, every second some lights would turn on and some would turn off. There were 15 lights numbered 1 to 15. They start all turned off. One second later all lights would turn on. Then all lights with an even number would turn off. One second later all lights with a number that is a multiple of 3 would switch (so if they were off, the are now on and vice-versa). One second later all lights with a number that is a multiple of 4 would switch. If this pattern continues, after 5 seconds how many lights are on?

1.4   ZIML January 2018 Division E

Below are the 20 Problems from the Division E ZIML Competition held in January 2018.
The answer key is available on p.161 in the Appendix.
Full solutions to these questions are available starting on p.106.

Problem 1

Natalie built a small castle using 12 cubes, as shown in the diagram

She plans on painting all the visible faces of her castle with glitter paint. How many faces will Natalie have to paint?

Problem 2

In the diagram below all triangles are equilateral, and the 4 smaller triangles in the middle are all of the same size.

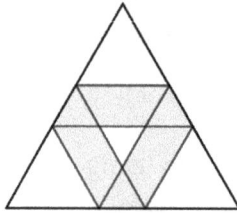

If the area of the shaded region is 48, what is the area of the whole triangle?

Problem 3

Kyle and Lisa each bought some candies. If Kyle gave 4 candies to Lisa, they would both have the same number of candies. They pooled their candies together and counted, and there were 24 in total. How many candies did Lisa have originally?

Problem 4

How many 2-digit numbers are divisible by their tens digit?

Problem 5

A sequence of numbers starts with

$$5, 7, 10, 14, 19, 25, 32, \ldots$$

If this pattern continues, what is the 12th number in the sequence?

Problem 6

Liam had 27 pints of chocolate chip and mint chocolate ice cream in his freezer. During the week he bought 3 pints of regular chocolate chip ice cream and ate 2 pints of mint chocolate chip ice cream. He has now three times as many pints of mint chocolate chip ice cream as regular chocolate chip ice cream in his freezer. How many pints of mint chocolate chip ice cream did he have before?

Problem 7

Rosemary loves cooking with fresh herbs so she has some pots with thyme and rosemary, which is her favorite, in her backyard. If she had 2 more thyme pots and 4 less rosemary pots, she would still have 10 more rosemary pots than thyme pots. If she has 24 pots in total, how many pots of rosemary does she have?

Problem 8

Two identical squares with side length 25 overlap to form a rectangle with area 1000, as shown on the diagram below.

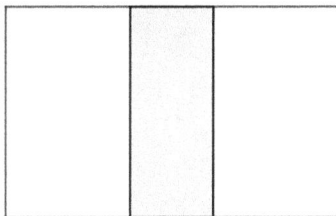

What is the area of the shaded region in the diagram?

Problem 9

How many different 4 digit numbers smaller than 6000 can be formed by rearranging the digits of 5706?

Problem 10

Find the greatest number smaller than 200 that leaves a reminder of 7 when divided by 15.

Problem 11

Rachel drew a square of area 16 square centimeters on a piece of paper. She then drew a square whose sides were 5 times as big as her first square. How many square centimeters is the area of the second square?

Problem 12

Randy paid $245.30 for a camera, including a 10% tax. What was the price of the camera before tax? Round your answer to the nearest cent.

Problem 13

Sophie and Betty drew similar triangles. Sophie's triangle is shown on the left and Betty's is shown on the right on the diagram below.

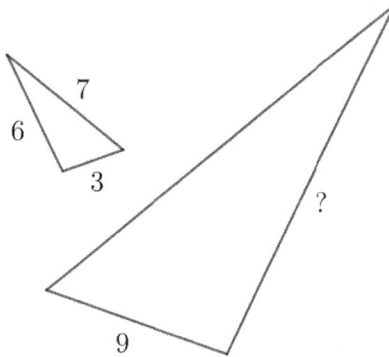

The two triangles have the same shape, but different position and size. What is the length of the missing side on Sophie's triangle?

Problem 14

The number 0.24 can be written as a fraction $\dfrac{P}{Q}$ in lowest terms. What is P?

Problem 15

Jerry drew 4 squares, each time doubling the length of their sides, and arranged them to form a spiral figure like in the diagram below.

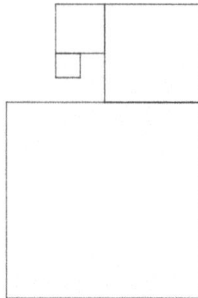

If the biggest square has area 256, what is the perimeter of the spiral?

Problem 16

Morgan needed 85 sticks for a project at school. Each stick is either 4 inches long or 7 inches long and the total length of all the sticks combined is 430 inches. How many 7 inch sticks are there?

Problem 17

How many numbers between less than 100 are multiples of 3 or 5, but not of both 3 and 5 at the same time?

Problem 18

A large bottle can hold 4 liters of oil, while every two small bottle can hold 1 liter of oil. A store has 100 liters of oil and the oil exactly fills up 60 bottles. How many large bottle does the store have?

Problem 19

Two pairs of congruent triangles are used to form a rectangle as shown on the diagram below.

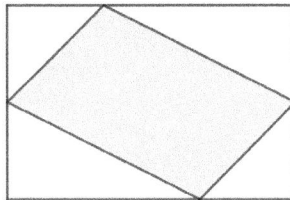

All triangles have the same height, and the length of the base of the smaller triangles is half of the length of the base of the bigger triangles. If the shaded region has area 72, what is the area of one of the smaller triangles?

Problem 20

You have two big buckets of water: one is red and the other one is blue. The red bucket contains 14 liters of water, and the blue bucket contains 18 liters of water. If you want bucket the red bucket to contain 3 times as much water as the blue bucket, how much water do you need to pour from the blue bucket into the red bucket?

1.5   ZIML February 2018 Division E

Below are the 20 Problems from the Division E ZIML Competition held in February 2018.
The answer key is available on p.162 in the Appendix.
Full solutions to these questions are available starting on p.114.

Problem 1
Using the digits 3, 7, 6, 0 and 2, exactly one time each, what is the smallest 5-digit even number we can write?

Problem 2
In the diagram, points A and B are the midpoints of their respective sides.

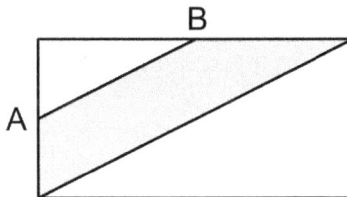

If the whole rectangle has an area of 24, what is the area of the shaded region? Round your answer to the nearest whole number if necessary.

Problem 3
Dustin and Gabe were eating muffins that had raisins in them. Neither of them like raisins, so they removed them from their muffins while they were eating them. Gabe found less raisins in his muffin and they noticed that the amount of raisins they found were in ratio 3 : 2. If they found 20 raisins altogether, how many raisins did Dustin find in his muffin?

Problem 4
At your school party, a big basket of sandwiches was delivered that weighs a total of 19 lbs. After half of the sandwiches were taken, the basket with the left-over sandwiches weighs 12 lbs. How much does the basket weigh in lbs?

Problem 5
What is the 15th term in the sequence 2, 3, 3, 4, 4, 4, 5, ... ?

Problem 6
Dante's dad was 24 when Dante was born. Three years ago his dad's age was three times his age. What is Dante's age today?

Problem 7
If we draw 2 different lines, they either intersect 0 times (if the lines are parallel) or they intersect 1 time. If we draw 4 different lines, what is the maximum number of times the lines intersect?

Problem 8
In a far off land there is a large community of dragons. One day, a group of 18 dragons of two different kinds were hanging out. Some of the dragons had 2 heads while the others had 3 heads. If there were 42 heads, how many 2-headed dragons were there?

Problem 9
Danny had an equilateral triangle, a square, and a regular pentagon, all with the same side length. She glued together the equilateral triangle and the square by one side, and the regular pentagon and the square by the opposite side she used for the triangle. She now had a new figure with many sides, all of the same size. How many sides does this new figure have?

Problem 10
Anton has a bag with 2 identical red balls and 3 identical green balls. If he grabs 3 balls all at once, how many different color combinations could he get? (The order Anton picks the balls does not matter.)

Problem 11
Juan took three quizzes for his History class. He scored a total of 228 points. The score in his first quiz was 8 points lower than his second quiz, and the score in his third quiz was 10 points lower than his second quiz. What was his score in the third quiz?

Problem 12

Sam and his brother Jack went out for dinner at the local diner. There were 8 different hamburgers and 6 soups available at the diner. Sam ordered first, and ordered one of the hamburgers or one of the soups. Jack ordered next, and while he ordered something different than Sam, he ordered the same type of food (so if Sam got a hamburger then Jack also got a hamburger). In how many different ways could Sam and Jack order their food?

Problem 13

In the following diagram the squares have side lengths 1, 2 and 3.

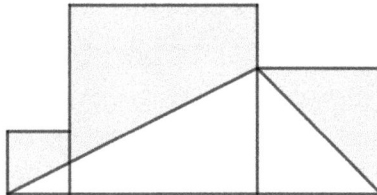

What is the shaded area? Round your answer to the nearest tenth if necessary.

Problem 14

George's grandparents visit every 10 days. His aunt and uncle visit every 14 days. Today both his grandparents and aunt and uncle are visiting. How many days will it be until the next time his grandparents and aunt and uncle visit on the same day?

Problem 15
Two friends leave the same place at the same time traveling in the same direction. One of them travels at a speed of 60 mph. After 3 hours the other friend is 18 miles behind. If they both travel at constant speed, what is the speed, in miles per hour, of the other friend? Round your answer to the nearest whole number if necessary.

Problem 16
Consider a list of 6 consecutive even numbers that have a sum of 78. What is the largest number in the list?

Problem 17
Larry just got a big fish tank for his restaurant. The fish tank is a rectangular prism that is 60 inches long, 30 inches wide, and 50 inches high. He needs to fill the tank so that the level of the water is 5 inches below the top edge of the tank. To fill the tank with water he is using a pitcher that can hold 180 cubic inches of water. How many times will he have to fill the pitcher to fill the tank with water?

Problem 18
Ruth has to pack dice that are $\frac{2}{3}$ inches long per side in boxes that are 12 inches long, 10 inches wide, and 5 inches high. Ruth packs the dice nicely, so that when they are stacked the square faces of the cubes line up. What is the maximum number of dice Ruth will be able to pack in each box?

Problem 19

Four congruent large rectangles and four congruent smaller rectangle are arranged to form a big square as in the diagram below.

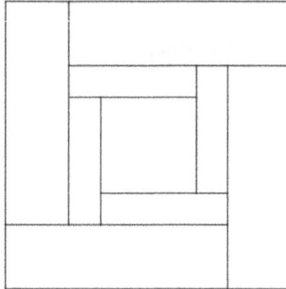

The width of the small rectangles is 1 and the width of the big rectangles is 2. If the area of the small square in the middle is 25, what is the area of the whole big square?

Problem 20

Adam and Bob played ping-pong. They agreed that whoever won 3 games first was the final winner. Assume that Bob was the final winner. In how many ways could the games have been played out?

1.6 ZIML March 2018 Division E

Below are the 20 Problems from the Division E ZIML Competition held in March 2018.
The answer key is available on p.163 in the Appendix.
Full solutions to these questions are available starting on p.121.

Problem 1
In a school in Jersey City there are 280 students that play hockey. If 60% of the students that play hockey are girls, how many boys play hockey?

Problem 2
Consider points A, B, C, and D on the number line.

If $AC = 7$, $AD = 13$ and $BD = 10$, what is BC?

Problem 3
What is the smallest odd number between 500 and 1000 that has no repeated digits and is a multiple of 5?

Problem 4
Mrs. Nicelady has 3 daughters and 2 sons. Each of her daughters has 3 daughters and 2 sons, and each of her sons have 2 daughters and 1 son. If each of Mrs. Nicelady's grandchildren have 2 children, how many descendants does Mrs. Nicelady have?

Problem 5

A sequence of numbers starts with $-2, 4, -6, 8, -10, \ldots$. If this pattern continues, what will be the 12th number on the list?

Problem 6

When gold sold for \$21 a troy ounce, Derek found \$84 of gold in his claim. The price of gold today is \$450 per troy pound. How many dollars is Derek's gold worth today? (One troy pound is 12 troy ounces.)

Problem 7

Tammara filled her backpack with berries she collected in a field. Her backpack together with the berries weighed 20 pounds. She decided to share some of her berries with her friends, and they all ate $\frac{2}{3}$ of the berries in the backpack. If the weight of the backpack is now 10 pounds, how many pounds does the backpack weigh when it is empty?

Problem 8

What is the difference between the largest 3-digit number and the smallest 3-digit number that can be formed using only the digits 3, 8, 2 and 9 if the numbers are not allowed to have repeated digits?

Problem 9

In the following diagram all of the squares have different side lengths.

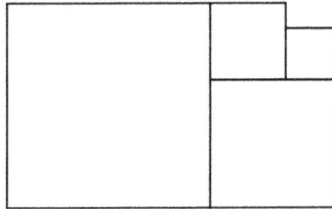

The smallest square has perimeter 12 and the second smallest has perimeter 20. What is the perimeter of the largest square?

Problem 10

Consider a rhombus inside a rectangle, creating 4 identical triangles as shown in the diagram below.

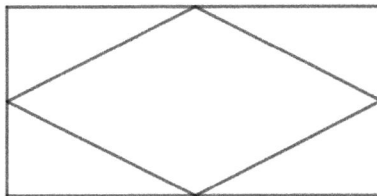

If the area of the rhombus is equal to 96, what is the area of one of the triangles?

Problem 11

A whole number bigger than 100 and smaller than 150 has the following properties: (i) it leaves no remainder when you divide it by 2; (ii) it leaves no remainder when you divide it by 5; and (iii) it leaves a remainder of 1 when you divide it by 7. What is the number?

Problem 12

A square has four times the area of a rectangle that has one side of length 2 and perimeter 20. What is the side length of the square?

Problem 13

Norton recently participated in a math competition where the exam had 25 questions. Each correct answer was worth 6 points, and for each incorrect or blank question 1 point was deducted from the final score. If Norton received 101 points, how many incorrect or blank questions did he have?

Problem 14

Duncan tried to draw a letter M in a grid made up with unit squares and he came up with the following figure.

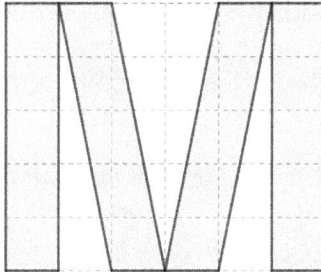

What is the area of the letter M?

Problem 15

Chris will buy some cakes for a big party. If he buys 4 cakes he would have $12 left. If he buys 7 cakes, he would need an extra $21 to pay for them. If all cakes cost the same, what is the price of one cake?

Problem 16

Ethan baked a cheese pie at home and decided to slice it and share it with this family. He knows how much his mom and dad love cheese pie, so he made sure to make a special slice for each of them that was twice as big as all other slices (which were all identical). In total Ethan cut the pie in 18 slices. If he measured the angle formed in one of the small pie slices, how many degrees would that angle be?

Problem 17

A train leaves a station, traveling at 70 miles per hour. Three hours later, a second train leaves on a parallel track, traveling on the same direction at 100 miles per hour. In how many hours will the second train catch up with the first train?

Problem 18

Fourteen identical rectangles are arranged to form a bigger rectangle as shown in the diagram below.

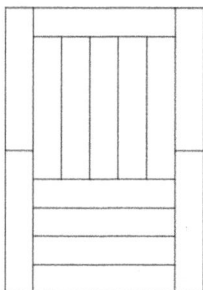

If the perimeter of the small rectangles is 60, what is the perimeter of the big rectangle?

Problem 19

David is afraid of Friday the 13th, particularly when it is in April. This year there is a Friday the 13th in April. In how many years will there be again a Friday the 13th in April? (Remember a normal year has 365 days and a leap year has one extra day. Leap years occur every 4 years and 2016 was a leap year.)

Problem 20

Alex is mixing soda flavors in the soda fountain. If he were to fill his cup just with grape flavored soda, it would take 30 seconds to fill his cup. If he were to fill his cup just with lime flavored soda, it would take 45 seconds to fill his cup. If he pours both grape and lime flavored soda at the same time, how many seconds would it take to fill his cup?

1.7   ZIML April 2018 Division E

Below are the 20 Problems from the Division E ZIML Competition held in April 2018.
The answer key is available on p.164 in the Appendix.
Full solutions to these questions are available starting on p.127.

Problem 1
Oliver was out collecting flowers. He collected 3 times as many red flowers as blue flowers and 2 times as many yellow flowers as blue flowers. Altogether Oliver collected 78 flowers. How many red flowers did Oliver collect?

Problem 2
Consider the figure in the diagram below.

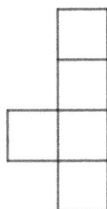

The shape is made out of 5 unit squares. In how many ways can one or two unit squares be attached to the figure so that the resulting figure has an axis of symmetry?

Problem 3
Grassy the grasshopper jumps jumps of length 3 inches or 5 inches. He never jumps the same length 3 times in a row. What is the smallest total length he can move by jumping 7 times?

Problem 4

Liam loves bananas and he usually eats 3 per day. Last week Liam ate 5 less bananas than this week, and this week he ate 2 more bananas than usual. How many bananas did Liam eat in total this week and last week?

Problem 5

How many square inches is the surface area of the triangular prism shown below?

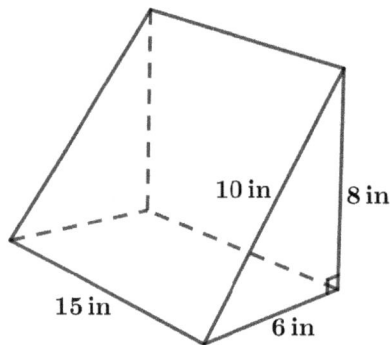

Problem 6

A sequence of numbers starts with $3, 2, 4, 1, 5, 0, 6, \ldots$. What is the next number in the sequence?

Problem 7

In the following diagram the shaded rectangle has two vertices on the midpoints of two of the sides of the trapezoid.

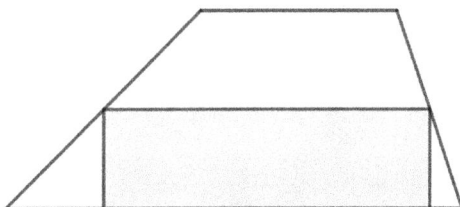

If the shaded rectangle has area 23, what is the area of the trapezoid?

Problem 8

Jamie has a triangle with integer side lengths and an area of 13.5 square inches. The base of her triangle is three times the height of her triangle. How many inches is the base of her triangle? Round your answer to the nearest tenth if necessary.

Problem 9

Mel is thinking of a number between 1 and 1000. If Mel swaps the last two digits of her number, it becomes 18 less than the original number. The hundreds digit in her number is half of its ones digit and one third of its tens digit. What number is Mel thinking of?

Problem 10

During a game show Ricky is asked 5 questions. He starts the game with 5 points and for every question he gets right he gets 1 point. If he does not answer or is incorrect he loses 1 point. After he finishes answering the questions he has 6 points. How many questions did he answer correctly?

Problem 11

Ms. Joy had a basket full of juicy apples, with the basket plus apples weighing a total of 21 pounds. She gave away $\frac{2}{3}$ of her apples (all apples had the same weight) and now her basket plus apples weighs 9 pounds. How many pounds of apples did Ms. Joy have originally?

Problem 12

Brendan and Kiara are interchanging secret messages using a number code. When sending their message, they substitute each digit separately. If Brendan wants to send the number 26510 to Kiara, he sends instead the number 57634 and Kiara will be able to decode the actual number using their code. If he wants to send the number 36978, he sends instead the number 97208 to Kiara. If Kiara received the number 82649, what was the number that Brendan actually sent?

Problem 13

The following diagram shows a square inscribed into a larger square.

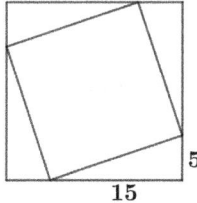

What is the area of the smaller square?

Problem 14

Dustin wrote the numbers from 150 to 365 on a piece of paper (starting with the number 150 and ending with the number 365). How many times did he write the digit 5?

Problem 15

Jovani is getting tired of people spelling his name wrong whenever he orders coffee at the local coffee shop. Over the past three months, 30% of the time they have spelled his name as Giovanni, 15 times they have spelled it Joevany, and 15% of the time they have spelled it as Jovan E .(the other times they spelled it correctly). If Jovani bought coffee 60 times, how many times did they spell his name correctly?

Problem 16

Natalie loves baking apple pies. For each pie she needs $2\frac{2}{3}$ cups of diced apples and $1\frac{1}{2}$ cups of all-purpose flour. After baking several pies she used 16 cups of diced apples. How many cups of all-purpose flour did she use to make the pies? Write your answer as a decimal rounded to the nearest tenth.

Problem 17

Patrick drew the blue regular hexagon shown below.

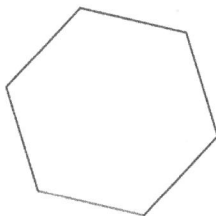

Then he drew green lines between opposite vertices of his hexagon, and finally he drew the remaining diagonals of his hexagon with red lines. Patrick noticed there were several red triangles in his final drawing. How many red triangles did Patrick see?

Problem 18

Bailey is training for a dancing competition. She knows staying hydrated is important, so she makes sure to drink 1 glass of water for every half hour of training. Bailey's glass can hold 250 milliliters of water and she refills it from a pitcher that can hold 1.5 liters of water. If Baileys pitcher is full, what is the longest time (in hours) she can train without refilling her pitcher? Round your answer to the nearest tenth if necessary.

Problem 19

How many numbers between 100 and 300 (including 100 and 300) are divisible by their hundreds digit?

Problem 20

Vic loves the rain, so whenever it starts raining he takes out a square bucket to collect water to play with later. Vic's bucket is in the shape of a cube that is 30 cm wide, 30 cm long, and 30 cm tall. Vic measured the height of the water after one hour, and two hours. After one hour the water reached a height of 6 cm, and after two hours it reached a height of 12 cm. If this rate continues, how many liters of water will there be in the bucket after 3 hours? Recall 1 liter is equal to 1000 cubic centimeters. Round your answer to the nearest tenth if necessary.

1.8   ZIML May 2018 Division E

Below are the 20 Problems from the Division E ZIML Competition held in May 2018.

The answer key is available on p.165 in the Appendix.

Full solutions to these questions are available starting on p.135.

Problem 1

What is the 10th number in the sequence $-5, 8, -11, 14, -17, \ldots$?

Problem 2

How many diagonals does a regular heptagon have? (Recall a heptagon has 7 sides.)

Problem 3

Curt is counting how many times he has to lick his lollipop to reach its center. He can lick his lollipop 4 times every 7 seconds. If it took him 637 seconds to reach the center, how many times did he have to lick his lollipop?

Problem 4

Six identical triangles with sides 3, 4, and 5 were used to create a figure like in the diagram below.

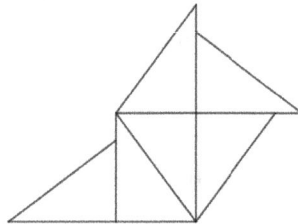

What is the perimeter of the figure?

Problem 5

Today in Celeste's class there are the same number of boys and girls. However, 4 girls and 3 boys did not make it to class since they caught the flu. If there are 8 boys in the class today, how many students are there in the class when everyone is present?

Problem 6

Beth wants to draw a rectangle with integer side lengths and a perimeter of 16. How many different rectangles can she draw? (Assume a rectangle with dimensions 2×3 is the same as a rectangle with dimensions 3×2.)

Problem 7

Howard needs to choose his clothes for a music festival. He will choose one of 5 pairs of shoes, one of 3 pairs of jeans, one of 5 shirts, and one of 3 hats. In how many different ways can Howard choose his outfit for the music festival?

Problem 8

The sum of the ages of Dan and Han is 56. 25 years ago Dan's age was 2 times Han's age. How old is Han today?

Problem 9

The Lexington family wants to wallpaper a room that is 12 feet long, 9 feet wide and 8 feet tall. The room has a door that is 1.5 feet wide and 7 feet tall, and a window that is 3 feet wide and 6 feet long and they do not wallpaper the ceiling (or the floor!). How many square feet of wallpaper are they going to use?

Problem 10

Daisy and some of her friends sit around a circle on the grass. They start counting off numbers, starting with the number 1. They stop when they reach the number 44. It turns out the same person that counted the number 2 also counted the number 44. If there is an odd number of kids sitting in the grass, at least 5 but no more than 15, how many kids are sitting on the grass?

Problem 11

Chris enjoys jumping long distances. He decided he would jump 4 times in a row and see how far he could get from where he was standing. Each time he jumped half as far as his previous jump. If his last jump was 0.5 meters long, how many meters did he jump in total?

Problem 12

Every week Duncan gets the same number of cookies on Sunday and eats the cookies (at least one cookie, but the same number each day) throughout the week from Sunday to the next Saturday. He then gives any leftover cookies to his sister. Duncan gets the same number of cookies every week, but the number of cookies he eats each day can change from week to week. Two weeks ago he gave his sister 4 cookies, last week he gave his sister 11 cookies, and this week he gave his sister 18 cookies. What is the minimum number of cookies Duncan gets every Sunday?

Problem 13

Corrine is making sandwiches for a picnic. She bought two loaves of bread, each with 20 slices, and four packs of sliced smoked turkey, each with 12 slices. If she wants to put 3 slices of smoked turkey in each sandwich, what is the maximum number of sandwiches she can make?

Problem 14

Rick needs to fill in a pool that is 8 feet wide, 20 feet long and 7 feet deep. How many gallons of water would he need, using the fact that 10 cubic feet of water is about the same as 75 gallons of water? Round your answer to the nearest gallon if necessary.

Problem 15

Lucas came across a machine that exchanged coins for small candies. The number of candies he gets will be such that its ones digit will be twice the number of pennies he uses, its tens digit will be the same as the number of dimes he uses, and its hundreds digit will be half of the number of quarters he uses. If Lucas inserts 4 pennies, 5 dimes, and 6 quarters, how many candies would he get?

Problem 16

The figure below shows three identical triangles inside of a square.

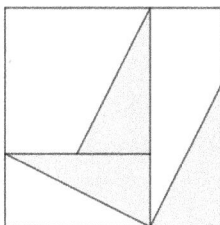

The length of the base of the triangles is half their height. If the side length of the square is 12, what is the shaded area?

Problem 17

Lindy has 20 hens and roosters at her farm. This morning she picked 33 eggs, and is sure each hen laid the same number of eggs. If there are more hens than roosters, how many roosters are there in Lindy's farm?

Problem 18
5 identical rectangles are arranged to form the figure below.

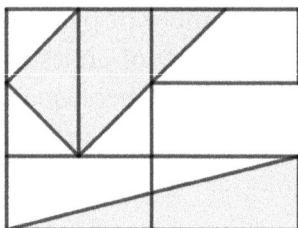

The shaded region has an area that is $N\%$ of the full figure. What is N?

Problem 19
Grace had a jar full of cookies. At first she ate 20% of the cookies in the jar. Then she ate 20% of the cookies left in the jar. If the jar had 50 cookies, how many cookies did Grace eat?

Problem 20
What is the greatest 2-digit number that leaves a remainder of 7 when you divide it by 12?

1.9 ZIML June 2018 Division E

Below are the 20 Problems from the Division E ZIML Competition held in June 2018.
The answer key is available on p.166 in the Appendix.
Full solutions to these questions are available starting on p.142.

Problem 1
Some identical cubes were arranged in the corner of a room as shown in the diagram below.

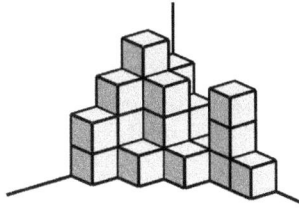

If there are no empty gaps between the wall and the cubes you can see, how many cubes were used in total?

Problem 2
Addie was asked to add several numbers for one of the questions in her math homework. She got 65536 as her answer, but then she noticed that she used the number 905 instead of the number 506. What was the correct answer?

Problem 3

Dustina needs to buy school supplies for her kids. She has 2 kids and the supply list says Dustina should buy, for each of her kids, 1 pair of scissors, 2 pencil erasers, 1 set of color pencils, 1 small ruler, 1 compass, 1 protractor, 10 pencils, 3 black pens, 2 red pens, 1 blue pen, and 1 small glue stick. Since her kids often lose supplies, she always buys one extra of each item on the list. How many items will Dustina buy in total?

Problem 4

How many lines of symmetry does the following figure have?

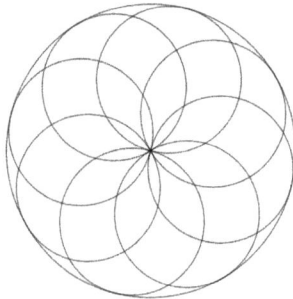

Problem 5

Daria just got a math card game. The rules of the game are simple: each player is dealt 3 cards with numbers on them and 2 cards with either $+$, $-$, $\times$, or $\div$ on them, then the player must arrange all 5 cards in a row to try to obtain the biggest number they can by performing the operations in the correct order. If Daria has cards with the numbers 8, 2, and 5, and operations $\times$ and $-$, what is the biggest number she could make?

Problem 6

At the warehouse there are bicycles and tricycles in storage. The manager of the warehouse counted 118 wheels in their last inventory. If there were 45 bicycles and tricycles in total, how many tricycles were there?

Problem 7

What is the smallest 3-digit number that leaves a remainder of 2 when it is divided by 3 or by 5.

Problem 8

Sam and Dan work in a flower shop making floral arrangements. Today they made 44 floral arrangements in 6 hours. Sam made 4 more than Dan. How many minutes, on average, did Sam spend making each of his arrangements? Round your answer to the nearest whole minute.

Problem 9

Mr. Don needs to fence a rectangular yard of area 75 square feet that is three times as long as it is wide. If Don needs to fence all four sides of the yard, how many feet of fence will he need?

Problem 10

One box of super small nails and one box of small nails weigh 8 pounds together. One box of small nails and one box of big nails weigh 18 pounds together. One box of super small and one box of big nails weigh 14 pounds together. How many pounds does one box of small nails weigh?

Problem 11

Several squares of different sizes are arranged in a line of length 13 as shown in the diagram below.

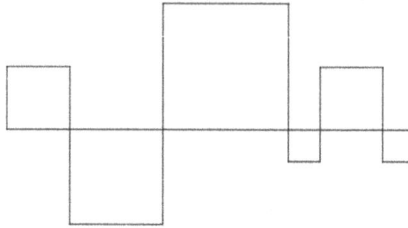

What is the sum of the perimeter of all squares?

Problem 12

Isabella has always been afraid of long words. She recently learned that the fear of long words is called Hippopotomonstrosesquipedaliophobia, which is itself a long word (with 35 letters). In order to overcome her fear, she decided to write repeatedly Hippopotomonstrosesquipedaliophobia in a piece of paper, with no spaces in between, until she filled the paper completely. Isabella filled 43 lines with 91 characters in each line. How many times did Isabella write the letter p? For reference, Hippopotomonstrosesquipedaliophobia contains 5 p's.

Problem 13

A rectangle is divided into 4 smaller rectangles with integer side lengths, as shown in the diagram below.

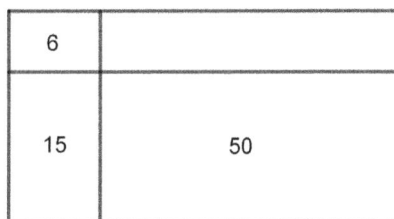

The areas of three of these rectangles are 6, 15, and 50. What is the area of the entire big rectangle?

Problem 14

Karina is milking the cows in her farm. She has 5 buckets of milk, each containing 15 liters. To sell the milk she will pour it into bottles that can hold half a gallon of milk. If one gallon is approximately 3.78 liters, how many bottles of of milk will she be able to fill completely?

Problem 15

As the end of the school year approaches, Rudy and 3 of his friends organized a gift exchange. Each of them will randomly pick a piece of paper with the name of the person they will buy a gift to. If anyone gets their own name, they start again so no one buys a gift for themselves and they have no clue of who is buying them a gift. In how many different ways could the gifts be exchanged?

Problem 16

Eusebio is building equilateral triangles with matchsticks, like the one in the diagram below, for which he used 18 matchsticks.

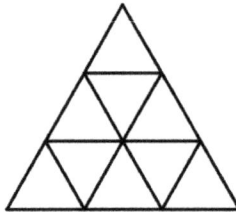

How many matches would he need to build a triangle with side length 7?

Problem 17

When Inez broke her piggy bank, she got 358 pennies, 172 quarters, 236 dimes, 15 silver dollars (each worth $1), and 50 nickels. She used her money to buy 3 movie tickets for her and her friends. After buying the movie tickets, she was left with $33.68. What was the cost of each movie ticket? Round your answer to the nearest cent.

Problem 18

What is the least common multiple of the first 10 even numbers?

Problem 19

The Daily Paper asked 200 of their readers to answer the following question: "If the rain was made out of chocolate, would you drink it?". 40% of the respondents were male, and 60% were female. Tobby was in charge of organizing the results in a table, but he did not finish his job. So far the table looks like:

	Yes	No	Don't know	Total
Male	?	6	14	?
Female	40	?	?	?
Total	?	?	70	200

According to the survey, what percent of the female respondents would not drink the rain if it was made out of chocolate?

Problem 20

You have available several copies of the figures shown in the diagram below.

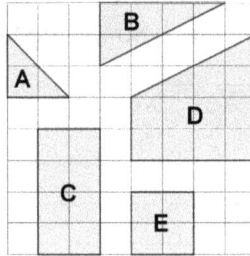

In how many different ways could you make a square using at most 4 figures? Note: it is OK to use the same figure more than once, as long as the total number of pieces used is at most 4. For example, two possible ways are (1) using only piece E forms a square or (2) using 2 of piece C.

2. ZIML Solutions

This part of the book contains the official solutions to the problems from the nine Division E ZIML Contests from the 2017-18 School Year.

Students are encouraged to discuss and share their own methods to the problems using the Discussion Forum on ziml.areteem.org.

2.1   ZIML October 2017 Division E

Below are the solutions from the Division E ZIML Competition held in October 2017.
The problems from the contest are available on p.17.

Problem 1 Solution
If one mouse is eating alone, it would take it 5 times as much time to eat a whole block of cheese, thus 1 mouse needs $5 \times 2 = 10$ hours to eat a whole block of cheese.

Answer: 10

Problem 2 Solution
Five years ago Lola was still 4 years older than Dola. As Lola's age was two times Dola's age, Dola was 4 years old and Lola was 8 years old. This means Lola is now $8 + 5 = 13$ years old.

Answer: 13

Problem 3 Solution
If Brandon had bought only type H pencils, he would have spent $2 \times 15 = 30$ dollars. That is $37 - 30 = 7$ less than what he actually spent. Since each HB pencil is $3 - 2 = 1$ dollar more expensive, he actually bought 7 type HB pencils and type H pencils.

Answer: 7

Problem 4 Solution
The shaded region is a trapezoid with height 10 and bases of length 4 and 14. Thus, the area of the shaded region is

$$\frac{10 \times (4 + 14)}{2} = 90.$$

Answer: 90

Problem 5 Solution

If the second die landed in 4, Drake could get the sums 5, 6, 7, 8, 9 and 10.

If second die landed in 5, he could get the sums 6, 7, 8, 9, 10 and 11.

If the second die landed in 6, he could get the sums 6, 7, 8, 9, 10 and 12.

So, the possible different sums he could get include all the numbers from 5 to 12, that is, there are 8 different possible sums.

Answer: 8

Problem 6 Solution

As the cakes spent on average 45 minutes in the oven, the oven was used for a total of $45 \times 2 = 90$ minutes. If the bigger cake had spent 10 minutes less in the oven, both cakes would have needed the same time and they would have used the oven for a total of $90 - 10 = 80$ minutes, thus the small cake was in the oven for $80 \div 2 = 40$ minutes.

Answer: 40

Problem 7 Solution

We can see the lines of symmetry using the diagram below.

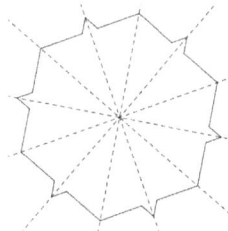

Therefore there are 6 lines of symmetry.

Answer: 6

Problem 8 Solution
From the first 7 terms we notice that the pattern is

$$\times 2, +2, \times 2, +2, \cdots$$

so the next 5 numbers in the sequence are

$$108, 110, 220, 222, 444$$

This means the 12th number in the sequence is 444.

Answer: 444

Problem 9 Solution
Note that 2 times the length of the rectangle is the same as 5 times its width. Since each rectangle has length 15, they must each have width $5 \times 2 \div 5 = 6$. The perimeter of the whole figure is then $6 \times 15 + 4 \times 6 = 114$.

Answer: 114

Problem 10 Solution
Since Mandy got a head start, we can pretend that the bag was missing 4 raisins, so it would have $124 - 4 = 120$ raisins total and Mandy ate exactly three times as many as Charly. This means Charly ate $120 \div (3 + 1) = 30$ chocolate covered raisins.

Answer: 30

Problem 11 Solution
Dorothy starts by placing the family picture in the 4th position. She then places the rest of the pictures one by one in each of the remaining 6 places on the wall. Thus, Dorothy can arrange the

pictures in $6 \times 5 \times 4 \times 3 \times 2 \times 1 = 720$ ways.

Answer: 720

Problem 12 Solution
The stairs consist of 6 rows of 4 cubes each, so Tyler used $6 \times 4 = 24$ cubes to build the whole model. Each cube has a volume of $2^3 = 8$ cubic centimeters, so the whole model has a volume of $24 \times 8 = 192$ cubic centimeters.

Answer: 192

Problem 13 Solution
If Gus can run 5 miles in one hour, then he can run $1.6 \times 5 = 8$ km in one hour. Since there are 60 minutes in each hour, to run 2 km Gus would need $2 \div 8 \times 60 = 15$ minutes.

Answer: 15

Problem 14 Solution
Note we can draw a big rectangle around the whole kite as shown in the diagram below.

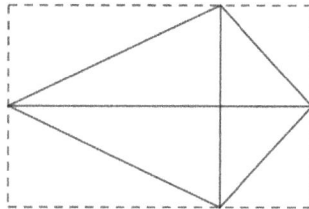

This rectangle has a length and width equal to the lengths of Katya's sticks, so the area of the rectangle is $30 \times 20 = 600$ square centimeters. As the kite itself makes up half of the rectangle, it has area $600 \div 2 = 300$ square centimeters.

Answer: 300

Problem 15 Solution

The multiples of 13 less than 100 are

$$13, 26, 39, 52, 65, 78, \text{ and } 91.$$

Adding 11 to each of these numbers we can find numbers that leave remainder 11 when dividing by 13:

$$24, 37, 50, 63, 76, 89, \text{ and } 102.$$

Therefore the largest 2-digit number that leaves remainder 11 when dividing by 13 is 89.

Answer: 89

Problem 16 Solution

Jerry will need two pieces of colored paper, each with an area of $30 \times 20 = 600$ square centimeters, for the bottom of the box. He will need four pieces of colored paper, each with area $30 \times 25 = 750$ square centimeters, to cover the front and back of the box. Lastly, he'll need four pieces of colored paper, each with area $20 \times 25 = 500$ square centimeters, to cover the sides of the box. Therefore Jerry needs

$$2 \times 600 + 4 \times 750 + 4 \times 500 = 6200$$

square centimeters of colored paper in total.

Answer: 6200

Problem 17 Solution

Every time Diego cut his pieces of paper in four equal pieces he got 4 times as many pieces of paper as before. Since he started with 2 pieces and he cut his pieces into fourths a total of 3 times, in the end he had

$$2 \times 4 \times 4 \times 4 = 2 \times 4^3 = 128$$

pieces of paper.

Answer: 128

Problem 18 Solution

The worst that could happen is that you grab all of the red and blue marbles, so, if you grab $4 + 6 = 10$ marbles or less, it could happen that you do not grab any of the green marbles. However, if you grab one more, you are guaranteed to have at least one green marble as there are only 10 blue and red marbles. Thus, you should grab 11 marbles to make sure you win that extra chocolate bar.

Answer: 11

Problem 19 Solution

Doris should bring enough pencils so that she can divide them evenly by both 5 (in case all of her friends show up) and by 3 (in case the twins don't show up). This means she must bring a number of pencils that is divisible by $3 \times 5 = 15$. The greatest multiple of 15 that is smaller than 117 is 105.

Answer: 105

Problem 20 Solution

Since each of the edges of the cube is 4 cm long and $23 \div 4 = 5$ with remainder 3, the ant visited 5 vertices after she started walking. Counting the vertex she started at, the ant visited a total of $5 + 1 = 6$ vertices.

Answer: 6

2.2   ZIML November 2017 Division E

Below are the solutions from the Division E ZIML Competition
held in November 2017.
The problems from the contest are available on p.25.

Problem 1 Solution
Each of the square tiles is 1 inch long per side. In the following
diagram we can see how they were arranged to form the figure.

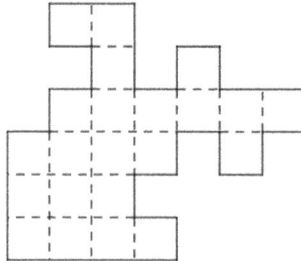

We can then count the number of edges in the perimeter of the
figure to find the perimeter of 34 inches.

Answer: 34

Problem 2 Solution
When the first new person enters the room, they shake hands with
12 people. The second person shakes hands with 13 people. The
third person will do with 14 people. The last person to enter the
room will shake hands with 15 people. Thus, $12 + 13 + 14 + 15 =$
54 handshakes occurred when the 4 people came into the room.

Answer: 54

Problem 3 Solution

Since the third and fifth tree are 70 feet apart, there are $70 \div 2 = 35$ feet between each two consecutive trees. Thus, the distance between the first and eight tree is $35 \times 7 = 245$ feet.

Answer: 245

Problem 4 Solution

Every 4 seconds $2 + 3 + 1 + 2 = 8$ drops fall to the sink. As there are $2 \times 60 = 120$ seconds in 2 minutes,

$$120 \div 4 \times 8 = 240$$

drops will fall to the sink in 2 minutes.

Answer: 240

Problem 5 Solution

Since both length and width of the rectangle are whole numbers, the possible dimensions of the rectangle are 1×42, 2×21, 3×14 and 6×7, each having perimeter 86, 44, 34, and 26, respectively. Thus, 86 is the largest possible perimeter the rectangle can have. Note the largest perimeter came from when the length and width were as far apart as possible.

Answer: 86

Problem 6 Solution

Every 1 hour Dylan runs $9 - 7 = 2$ more miles than Ala. So, after 1.5 hours they are $2 \times 1.5 = 3$ miles apart from each other.

Answer: 3

Problem 7 Solution

The left and right sides of the parallelogram each have length 2 cm, so the remaining sides of the parallelogram have length 40 cm. As each triangle has a side length of 2 cm, there are

$$40 \div 2 = 20$$

triangles in total.

Answer: 20

Problem 8 Solution

Since 5 and 50 are said by the same student, the number of students is a factor of $50 - 5 = 45$. 45 has factors $1, 3, 5, 9, 15, 45$, so there must have been 15 friends in total.

Answer: 15

Problem 9 Solution

Since Casey is painting 300 dimples, it will take him $300 \times 2 = 600$ seconds to paint all of them. Since there are 60 seconds in one minute, it will take him $600 \div 60 = 10$ minutes in total.

Answer: 10

Problem 10 Solution

Since every day he added twice as many crunches as the day before, the fourth day he will add $2 \times 4 = 8$ for a total of $16 + 8 = 24$, and the fifth day he will add $2 \times 8 = 16$ for a total of $24 + 16 = 40$ crunches.

Answer: 40

Problem 11 Solution

Let's pretend that no ghosts left the party around midnight. So, when the 5 werewolves arrived, there were $37 + 5 = 42$ monsters at the party and there were $4 - 2 = 2$ more ghosts than werewolves.

Had there been 2 more werewolves, there would have been the same number of each monster and a total of $42 + 2 = 44$ monsters, so there were $44 \div 2 = 22$ ghosts.

Answer: 22

Problem 12 Solution
In the diagram below we can see there are 12 lines of symmetry.

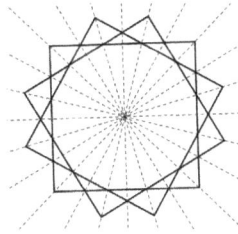

Note there are 6 lines of symmetry through opposite vertices of the squares and then 6 additional lines of symmetry where the squares intersect.

Answer: 12

Problem 13 Solution
Cindy's entire candy bar has area

$$4 \times 8 = 32$$

square inches. Thus half the candy bar is $32 \div 2 = 16$ square inches. Luke's candy bar has area

$$2 \times 10 = 20,$$

so he must eat

$$\frac{16}{20} = \frac{4}{5}$$

of his candy bar. We have

$$\frac{4}{5} = 80\%$$

so our answer is 80.

Answer: 80

Problem 14 Solution

First, we need to find the total distance the truck travels, which is

$$50 + 210 + 100 = 360 \text{ miles.}$$

Then the total hours the truck traveled is

$$1 + 3 + 2 = 6 \text{ hours.}$$

hours. So, the average speed of the truck is

$$360 \div 6 = 60 \text{ miles per hour.}$$

Answer: 60

Problem 15 Solution

Folded up the tetrahedron looks like

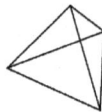

From the diagram we see it has 6 edges.

Answer: 6

Problem 16 Solution

Note that
$$3030303 \times 5050505 \text{ and } 303 \times 505$$

will have the same last three digits. We have $303 \times 505 = 153015$, so $A = 0$ and $B = 1$. Therefore, $A + B = 1$.

Answer: 1

Problem 17 Solution

First we convert minutes to hours. Since there are 60 minutes in 1 hours, we have that

$$90 \text{ minutes} = \frac{90}{60} = \frac{3}{2} \text{ hours.}$$

Similarly
$$30 \text{ minutes} = \frac{30}{60} = \frac{1}{2} \text{ hours.}$$

Therefore, in total Melissa travels

$$\frac{3}{2} \times 50 + \frac{1}{2} \times 60 = 75 + 30 = 105$$

miles.

Answer: 105

Problem 18 Solution

If we want the number to be divisible by 5, it has to end in 5. The largest 4-digit number using only the digits 3 and 5 is 5555, but $5555 = 1851 \times 3 + 2$, so is not divisible by 3. The next largest number (remember the last digit still needs to be 5) is 5535, and since $5535 = 1845 \times 3$, 5535 is divisible by 3 and 5 as needed.

Alternatively, a number is divisible by 3 only if the sum of its digits is also divisible by 3, so we must use the digit 5 in our number three times. The largest 5-digit number we can make using the digit 5 three times and the digit 3 once is 5535.

Answer: 5535

Problem 19 Solution

Robert's pool has a volume of

$$15 \times 20 \times 7 = 2100$$

cubic feet. Therefore Robert's pool holds about

$$2100 \times \frac{75}{10} = 210 \times 75 = 15750$$

gallons of water.

Answer: 15750

Problem 20 Solution

In 1 minute the hot water tap fills $\frac{1}{20}$ of the bathtub, the cold water tap fills $\frac{1}{15}$ of the bathtub, and the $\frac{1}{30}$ of the bathtub goes through the drain. So after 1 minute

$$\frac{1}{20} + \frac{1}{15} - \frac{1}{30} = \frac{1}{12}$$

of the bathtub is full. This means the bathtub would be completely full in 12 minutes.

Answer: 12

2.3 ZIML December 2017 Division E

Below are the solutions from the Division E ZIML Competition held in December 2017.
The problems from the contest are available on p.33.

Problem 1 Solution
The cows in total eat $3 \times 16 = 48$ bundles of hay per day. Similarly the horses eat $4 \times 14 = 56$ bundles of hay per day. Hence in total they eat $48 + 56 = 104$ bundles of hay in a single day. Since there are 7 days in a week, the farmer needs $7 \times 104 = 728$ bundles of hay each week.

Answer: 728

Problem 2 Solution
Work with one digit at a time: the ones digit is $2 \times 2 = 4$; the hundreds digit is $2 \times 4 = 8$. The number is 824.

Answer: 824

Problem 3 Solution
Divide the figure into 4 identical triangles as in the diagram below.

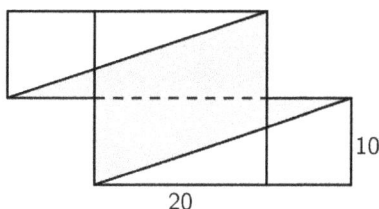

As 2 of the 4 triangles are shaded, we see the shaded region is

$$\frac{2}{4} = \frac{1}{2} = 50\%$$

of the full figure.

Answer: 50

Problem 4 Solution

The number that he keeps repeating has 10 digits in total. So, every 10th digit will be a 7, always followed by a 1. Since $167 \div 10 = 16$ with remainder 7, the last number Anton wrote will be the 7th number in the original 10-digit number, so 6.

Answer: 6

Problem 5 Solution

Grouping we have

$$(-5 + 10) + (-15 + 20) + \cdots + (-195 + 200) = 20 \times 5 = 100.$$

Answer: 100

Problem 6 Solution

The ratio of jars of honey to fish is $3 : 5$, and the ratio of fish to potatoes is $2 : 6 = 1 : 3$. Since $1 : 3 = 5 : 15$, we can trade 35 potatoes for 5 fish, and then use those to trade for 3 jars of honey. This means the ratio of jars of honey to potatoes is $3 : 15 = 1 : 5$. Thus, one jar of honey is worth 5 potatoes.

Answer: 5

Problem 7 Solution

Connect A to C and B to D. Note this divides the rectangle into 8 triangles that are all the same. Since half of these triangles are shaded, the shaded region has area $30 \div 2 = 15$.

Answer: 15

Problem 8 Solution

If Patrick had gotten 20 points less in the math test and 10 points more in the history score, he would have had $250 - 20 + 10 = 240$ points in total and he would have had the same score in all exams. This means he would have had a score of $240 \div 3 = 80$ in all exams. Therefore, he actually got $80 + 20 = 100$ points in math exam.

Answer: 100

Problem 9 Solution

Since 40% of the students who play volleyball are boys, the other 60% are girls. Hence the number of girls who play volleyball is

$$60\% \times 180 = \frac{6}{10} \times 180 = 6 \times 18 = 108.$$

Answer: 108

Problem 10 Solution

For Adam's square to have area 25, we see the side length must be 5 (as $5 \times 5 = 25$). This is the height of Bob's triangle, so if Bob's triangle has area 7.5, the base must have length 3 (as $\frac{1}{2} \times 3 \times 5 = 7.5$). Therefore, Cynthia's square has side length 3, so area $3 \times 3 = 9$.

Answer: 9

Problem 11 Solution

You will use the least amount of bills if you use the largest bills first. So, you need to use 1 $500 bill, 1 $100 bill, 2 $20 bills, 2 $5 bills, and 4 $1 bills. Thus, you will need to use a total of $1 + 1 + 2 + 2 + 4 = 10$ bills.

Answer: 10

Problem 12 Solution

The width and length of each of the small rectangles are in ratio $1 : 3$. As the perimeter of each small rectangle is 16, they have width 2 and length 6. Thus, the big rectangle has width $2 + 6 = 8$ and length $6 + 6 = 12$. The perimeter of the big rectangle is then $8 + 8 + 12 + 12 = 40$.

Answer: 40

Problem 13 Solution

In total, you and your brother have $103 + 92 = 195$ marbles. Since now your brother has 4 times as many marbles as you, that means you have $\frac{1}{5}$ of the total number of marbles and your brother has the rest. This means you actually have $195 \div 5 = 39$, which tells us you lost $103 - 39 = 64$ marbles.

Answer: 64

Problem 14 Solution

The location of Frank and Lacey is marked with an $\times$ below, with the ice cream shops also shown.

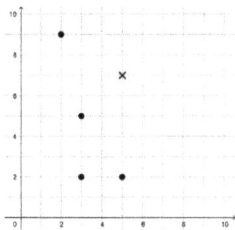

From this diagram we see that the ice cream shop at $(3, 5)$ is the closest. Therefore the sum of the x and y-coordinates is $3 + 5 = 8$.

Answer: 8

Problem 15 Solution

You need to at least color the extra squares marked below.

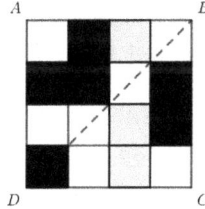

Hence 3 squares must be colored.

Answer: 3

Problem 16 Solution

We know that they need to gather 24 insects in total, since they need to use all of the 24 hats and each insect has only one head. If Greta were to gather 24 dimplies, she would use

$$3 \times 24 = 72$$

shoes, which means they would still have

$$106 - 72 = 34$$

shoes left. Since they must use all shoes, they need at least some rumplies. A rumply has $5 - 3 = 2$ more feet than a dimply. Since they need to use those 34 extra shoes, this means Hans should gather instead $34 \div 2 = 17$ rumplies and Greta should gather $24 - 17 = 7$ dimplies.

Answer: 17

Problem 17 Solution

We know either Bonnie or Carlo drives. If Bonnie drives, then Ali must sit in the front passenger seat and Carlo and Dianna sit in the back seats. Either Carlo sits on the left and Dianna on the right, or Dianna on the left and Carlo on the right. Hence there are 2 seating arrangements if Bonnie drives:

$$B \quad A \quad \text{or} \quad B \quad A$$
$$C \quad D \qquad \quad D \quad C$$

If Carlo drives, then Bonnie and Ali must sit in the back seats and Dianna in the front passenger seat. This again gives 2 more seating arrangements:

$$C \quad D \quad \text{or} \quad C \quad D$$
$$A \quad B \qquad \quad B \quad A$$

Hence there are $2 + 2 = 4$ total seating arrangements.

Answer: 4

Problem 18 Solution

All squares are rectangles, so we are looking for the number of rectangles that are not squares. Let's count first the horizontal rectangles: we have 9 1×2 rectangles, 6 1×3 rectangles, 3 1×4 rectangles, 4 2×3 rectangles, 2 2×4 rectangles, and 1 3×4 rectangle. Now let's count the vertical rectangles: we have 8 1×2 rectangles, 4 1×3 rectangles, and 3 2×3 rectangles. Thus, in total we have $9 + 6 + 3 + 4 + 2 + 2 + 1 + 8 + 4 + 3 = 40$ rectangles that are not squares.

Answer: 40

Problem 19 Solution

If your ticket number ends in 3, there are 18 tickets you can buy:

$$003, 013, 023, 043, \ldots, 103, 113, 123, 143, \ldots 193.$$

Similarly, if the second digit of your ticket number is 3, there are another 18 tickets you can buy:

$$030, 031, 032, 034, \ldots, 130, 131, 132, 134, \ldots 139.$$

Hence, there are a total of 36 tickets that you can buy.

Answer: 36

Problem 20 Solution

In the table below you can see the pattern of the lights during the first 5 seconds. $\times$ means the light is turned off and $\circ$ means it is turned on.

	0 sec	1 sec	2 sec	3 sec	4 sec	5 sec
1	×	○	○	○	○	○
2	×	○	×	×	×	×
3	×	○	○	×	×	×
4	×	○	×	×	○	○
5	×	○	○	○	○	×
6	×	○	×	○	○	○
7	×	○	○	○	○	○
8	×	○	×	×	○	○
9	×	○	○	×	×	×
10	×	○	×	×	×	○
11	×	○	○	○	○	○
12	×	○	×	○	×	×
13	×	○	○	○	○	○
14	×	○	×	×	×	×
15	×	○	○	×	×	○

We can see that after 5 seconds there are 9 lights turned on.

Answer: 9

2.4 ZIML January 2018 Division E

Below are the solutions from the Division E ZIML Competition held in January 2018.

The problems from the contest are available on p.41.

Problem 1 Solution

Each of the four top corner cubes have 5 exposed faces, each of the bottom corner cubes have 2 exposed faces, and each of the other 4 cubes have 3 exposed faces. Thus, Natalie will need to paint

$$4 \times 5 + 4 \times 2 + 4 \times 3 = 40$$

cube faces.

Answer: 40

Problem 2 Solution

Note the side length of the medium sized equilateral triangles is twice the side length of the smaller triangles, thus, we can divide the whole triangle in smaller triangles like in the following diagram:

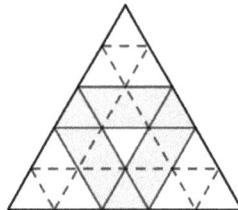

We can see then the shaded area is made up of 12 smaller triangles, and the whole triangle is made up of 25 smaller triangles. The area of one smaller triangle is then $48 \div 12 = 4$, so the area of the bigger triangle is $4 \times 25 = 100$.

Answer: 100

Problem 3 Solution
Kyle has $4 + 4 = 8$ more candies than Lisa. Kyle has more candies, so he must have

$$(24 + 8) \div 2 = 16$$

pieces of candy and Lisa $16 - 8 = 8$ pieces of candy.

Answer: 8

Problem 4 Solution
First note that any 2-digit number that ends in 0 can be divided by its tens digit. To find the rest of the numbers, we can keep adding the tens digit until we do not have the same tens digit anymore. So, all the 2-digit numbers that can be divided by their tens digit are

$$10, 11, 12, 13, 14, 15, 16, 17, 18, 19, 20, 22, 24, 26, 28,$$

$$30, 33, 36, 39, 40, 44, 48, 50, 55, 60, 66, 70, 77, 80, 88, 90, 99,$$

so there are 32 of them.

Answer: 32

Problem 5 Solution
The pattern we can see from the first few numbers in the list is $+2, +3, +4 \ldots$, so the next few numbers on the sequence are

$$40, 49, 59, 70, 82 \ldots$$

The 12th number in the sequence is 82.

Answer: 82

Problem 6 Solution

After he bought 3 pints of regular chocolate chip and ate 2 pints of mint chocolate chip he had a total of $27 + 3 - 2 = 28$ pints of ice cream. Of those, for every pint of regular chocolate chip ice cream he has 3 pints of mint chocolate chip ice cream, so he has $28 \div 4 = 7$ pints of regular chocolate chip and $7 \times 3 = 21$ pints of mint chocolate chip ice cream. This means he had $21 + 2 = 23$ pints of mint chocolate chip ice cream.

Answer: 23

Problem 7 Solution

Let's pretend she indeed has 2 more thyme pots and 4 less rosemary pots. This way she has $24 + 2 - 4 = 22$ pots in total and she still has 10 more rosemary pots than thyme pots. This means she would have

$$(22 + 10) \div 2 = 16$$

rosemary pots and $16 - 10 = 6$ thyme pots. Since we pretended we had 2 more thyme pots and 4 less rosemary pots, Rosemary actually has $16 + 4 = 20$ rosemary pots and $6 - 2 = 4$ thyme pots.

Answer: 20

Problem 8 Solution

The big rectangle has the same width as the side length of the two squares, and since it has area 1000, the length of the rectangle is $1000 \div 25 = 40$. The shaded rectangle in the diagram has thus length 25 and width $25 + 25 - 40 = 10$. Therefore, it has area $25 \times 10 = 250$.

Answer: 250

Problem 9 Solution

The first digit of the number must be less than 6 and not 0, so the only option is for the number to start with 5. We can then arrange the other three digits in any order, we have 3 choices for the second digit, 2 choices for the third digit and 1 choice for the last digit. So, there are $3 \times 2 \times 1 = 6$ numbers less than 6000 that can be obtained by rearranging the digits of 5706.

Answer: 6

Problem 10 Solution

A number that leaves a remainder of 7 when we divide by 15 is 7 more than a multiple of 15. The greatest number that is a multiple of 15 and less than 200 is 195, but $195 + 7 = 202$ is greater than 200. The second greatest number that is a multiple of 15 less than 200 is 180, so the number we are looking for is $180 + 7 = 187$.

Answer: 187

Problem 11 Solution

The first square that Rachel drew was 4 cm long per on each side. Thus, the second square she drew had sides of length $4 \times 5 = 20$ cm. So, the area of the second square is $20 \times 20 = 400$ square centimeters.

Answer: 400

Problem 12 Solution

The price that includes tax is equal to $100 + 10 = 110$ percent of the original price. Thus, the price of the camera with no tax is $245.30 \div 110\% = 245.30 \div 1.10 = 223.00$ dollars.

Answer: 223

Problem 13 Solution

As the triangles have the same shape, they are similar, so their sides are proportional. We can see the smaller side of Sophie's triangle is 3 units long, and the smaller side of Betty's triangle is 9 units long, so the sides of Betty's triangle are 3 times as big as those of Sophie's. The length of the missing side on Betty's triangle is then $6 \times 3 = 18$.

Answer: 18

Problem 14 Solution

As a fraction, 0.24 is the same as $\dfrac{24}{100}$. We can divide both 24 and 100 by 4, so we can simplify the fraction to $\dfrac{6}{25}$. Thus, $P = 6$.

Answer: 6

Problem 15 Solution

If the area of the big square is 256, that means it has side length 16, as $16 \times 16 = 256$. Thus, the smaller squares have side lengths 8, 4 and 2. The perimeter of the spiral is then

$$2 + 2 + 2 + 4 + 4 + 8 + 8 + 16 + 16 + 16 + 8 + 4 + 2 = 92.$$

Answer: 92

Problem 16 Solution

If all the sticks were 4 inch sticks, the total length of the 85 sticks would be

$$85 \times 4 = 340$$

inches. Since the actual length is

$$430 - 340 = 90$$

inches longer, some of the 4 inch sticks need to be replaced with 7 inch sticks. Each 7 inch stick is

$$7 - 4 = 3$$

inches longer. Since the actual length is 90 more than the length of all 4 inch sticks, we must replace

$$90 \div 3 = 30$$

4 inch sticks with 7 inch sticks. Hence there are

$$85 - 30 = 55$$

four inch sticks and 30 seven inch sticks.

Answer: 30

Problem 17 Solution
A number that is a multiple of both 3 and 5 if it is a multiple of 15. Note $99 \div 3 = 33$, $99 \div 5 = 19$ with remainder 4, and $99 \div 15 = 6$ with remainder 9, so there are 33 multiples of 3, 19 multiples of 5 and 6 multiples of both 3 and 5. Therefore, there are

$$33 + 19 - 6 = 46$$

numbers less than 100 that are multiples of 3 or 5, but not both.

Answer: 46

Problem 18 Solution
Since two small bottles hold 1 liter of oil, each small bottle holds

$$1 \div 2 = 0.5$$

liters of oil. If the store uses only small bottles, it can only hold

$$60 \times 0.5 = 30$$

liters of oil. This is

$$100 - 30 = 70$$

less than they need to hold. Each large bottle holds

$$4 - 0.5 = 3.5$$

extra liters of oil. Hence if they switch

$$70 \div 3.5 = 20$$

bottles from small to large they can hold the correct amount of oil with 60 bottles. Therefore, there are

$$60 - 20 = 40$$

small bottles and 20 large bottles.

Answer: 20

Problem 19 Solution

Note we can split the shaded region into 4 triangles (using 2 triangles of each kind).

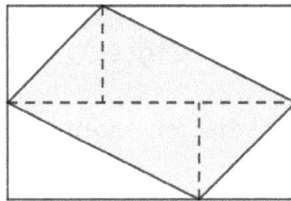

Since the base of the bigger triangles is twice the base of the smaller triangles, the area of two small triangles is the same as the area of a big triangle. This means the shaded area is the same as $2 + 1 + 2 + 1 = 6$ small triangles. Thus, the area of a small triangle is $72 \div 6 = 12$.

Answer: 12

Problem 20 Solution

We have in total $14 + 18 = 32$ liters of water. We want the red bucket to have 3 times as much water as the blue bucket, so for every 1 liter of water that we blue bucket has, we want the red bucket to have 3 liters. This means we want the blue bucket to have $32 \div (3 + 1) = 8$ liters of water. Right now the blue bucket has 18 liters of water, so we need to pour $18 - 8 = 10$ liters of water from the blue bucket to the red bucket.

Answer: 10

2.5 ZIML February 2018 Division E

Below are the solutions from the Division E ZIML Competition held in February 2018.

The problems from the contest are available on p.49.

Problem 1 Solution

The number has to start with a digit that is not 0, and must end with either 0, 2 or 6. To get the smallest number we shall use the smallest digits first, thus the smallest number we can make is 20376.

Answer: 20376

Problem 2 Solution

Connect the other midpoints as in the diagram below.

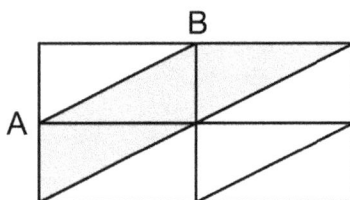

There are 8 total triangles, with 3 of them shaded. Hence the area of the shaded region is $\frac{3}{8} \times 24 = 9$.

Answer: 9

Problem 3 Solution

As the number of raisins they found is in ratio $3 : 2$, for every $3 + 2 = 5$ raisins they found together, 3 of them were Dustin's and 2 of them were Gabe's. That is, Dustin found $20 \div 5 \times 3 = 12$

raisins in his muffin.

Answer: 12

Problem 4 Solution

The basket and sandwich weight decreased by $19 - 12 = 7$ lbs when half of the sandwiches were eaten. Thus in total the sandwiches weigh $2 \times 7 = 14$ lbs. The remaining $19 - 14 = 5$ lbs must be the weight of the basket.

Answer: 5

Problem 5 Solution

Note 2 appears 1 time, followed by 3 two times, and 4 three times. We should have then 5 four times, 6 five times, 7 six times, and so on. Thus, the first 15 terms of the sequence are

$$2, 3, 3, 4, 4, 4, 5, 5, 5, 5, 6, 6, 6, 6, 6, \ldots$$

and the 15th term of the sequence is 6.

Answer: 6

Problem 6 Solution

Three years ago the ratio of the ages of Dante and his dad was $3 : 1$. His dad is is 24 years older than him, so three years ago Dante was 12 and his dad was 36. Thus, Dante is $12 + 3 = 15$ years old today.

Answer: 15

Problem 7 Solution

Start by drawing two lines. They intersect at 1 point. Draw a third line that intersects both lines, so we have 2 more intersection points. Draw a fourth line that intersects all three other lines, so we have 3 more intersection points. In total, we have $1 + 2 + 3 = 6$

intersection points.

Answer: 6

Problem 8 Solution

If all 18 dragons had been 3-headed dragons, there would have been $3 \times 18 = 54$ heads. That is $54 - 42 = 12$ more heads than there actually were. As a 3-headed dragon has $3 - 2 = 1$ more head than a 2-headed dragon, there were 12 2-headed dragons.

Answer: 12

Problem 9 Solution

The equilateral triangle has 3 sides, the square has 4 sides and the regular pentagon has 5 sides. When she glues together the triangle and the square she obtains a figure that has $3 + 4 - 2 = 5$ sides, as now the triangle and the square share one side. Similarly, when she glues the pentagon to another side of the square, she obtains a figure with $5 + 5 - 2 = 8$ sides. The resulting figure looks like

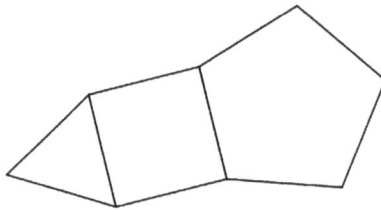

Answer: 8

Problem 10 Solution

He could get all 3 of them green, 2 green and 1 red, or 1 green and 2 red, so there are 3 different color combinations he could get.

Answer: 3

Problem 11 Solution

If Juan had scored 8 more points on his first quiz, and 10 more points on his third quiz, he would have had $228 + 10 + 8 = 246$ points in total and the same score in all three quizzes. That means he scored $246 \div 3 = 82$ points on his second quiz and thus, $82 - 10 = 72$ points on his third quiz.

Answer: 72

Problem 12 Solution

If they both order burgers then Sam has 8 choices and Jack has 7 choices. If they decide to order soup, Jack has 6 choices, and Sam has 5 choices. Thus, in total they have $8 \times 7 + 6 \times 5 = 86$ different ways to order their food.

Answer: 86

Problem 13 Solution

The shaded area is equal to the sum of the areas of the squares minus the area of the unshaded triangle. The triangle has base $1 + 2 + 3 = 6$ and height 2, so it has area $6 \times 2 \div 2 = 6$. Thus, the shaded area is

$$1 \times 1 + 2 \times 2 + 3 \times 3 - 6 = 8.$$

Answer: 8

Problem 14 Solution

We are looking for the least common multiple of 10 and 14. 10 is a multiple of 2 and 5. 14 is a multiple of 2 and 7. Therefore we want a multiple of 2, 5, and 7. The smallest such number is 70.

Answer: 70

Problem 15 Solution

The second friend must be going slower, since he is 18 miles behind. As they are traveling at constant speed, every hour the second friend gets behind by $18 \div 3 = 6$ miles. That means his speed is 6 less miles per hour than his friend. Thus, the speed of the other friend is $60 - 6 = 54$ mph.

Answer: 54

Problem 16 Solution

The first number on the list is $5 \times 2 = 10$ less than the last number on the list. Similarly the second, third, fourth and fifth numbers are, respectively, 8, 6, 4 and 2 less than the last number on the list. If we increase the first number by 10, the second by 8, and so on, all numbers will be as big as the last number on the list and the sum of all numbers will be

$$78 + 10 + 8 + 6 + 4 + 2 = 108.$$

Thus, the largest number on the list is $108 \div 6 = 18$.

Answer: 18

Problem 17 Solution

As he only needs the water to be up to a height of $50 - 5 = 45$ inches, Larry will need $60 \times 30 \times 45$ cubic inches of water. Thus, he will need to fill and pour the pitcher

$$\frac{60 \times 30 \times 45}{180} = 10 \times 45 = 450$$

times in total.

Answer: 450

Problem 18 Solution

Dividing each dimension of the box by the length of the side of one die, we will find out how many dice fit along, across, and high. The box can fit $12 \div \frac{2}{3} = 18$ dice along, $10 \div \frac{2}{3} = \frac{30}{2} = 15$ dice across, and $5 \div \frac{2}{3} = \frac{15}{2} \approx 7$ dice high. Therefore, each box can fit a maximum of $18 \times 15 \times 7 = 1890$ dice.

Answer: 1890

Problem 19 Solution

As the area of the small square is 25, it has side length 5. This means the smaller rectangles have length $5 + 1 = 6$. Thus, the side length of the big square is $2 + 6 + 1 + 2 = 11$. Therefore, the area of the big square is 121.

Answer: 121

Problem 20 Solution

We will use brute force to determine the various outcomes of the games. Given that Bob is the final winner, the score of the ping-pong match could either be 3-0, 3-1, or 3-2.

If the final score is 3-0, the possibility is

BBB.

There is only 1 possibility.

If the final score is 3-1, the possibilities are

ABBB, BABB, BBAB.

There are 3 possibilities.

If the final score is 3-2, the possibilities are

$$AABBB, ABABB, ABBAB, BAABB, BABAB, BBAAB.$$

There are 6 possibilities.

Therefore, there are
$$6 + 3 + 1 = 10$$
total ways that the game could have been played out.

Answer: 10

2.6 ZIML March 2018 Division E

Below are the solutions from the Division E ZIML Competition held in March 2018.

The problems from the contest are available on p.55.

Problem 1 Solution

Since 60% of the students that play hockey are girls, the remaining $100\% - 60\% = 40\%$ percent are boys. Hence, the number of boys that play hockey is $280 \times 40\% = 280 \times 0.4 = 112$.

Answer: 112

Problem 2 Solution

Since $AD = 13$ and $AC = 7$, $CD = 13 - 7 = 6$.

Therefore $BC = 10 - 6 = 4$.

Answer: 4

Problem 3 Solution

Since the number is divisible by 5, it must end in 0 or 5, but since it is odd, it can't end in 0. As the number has different digits, the first digit will be 6, so the number is 605.

Answer: 605

Problem 4 Solution

Each of Mrs. Nicelady's daughters has $3 + 2 = 5$ children, and each of her sons has $2 + 1 = 3$ children, so she has

$$3 \times 5 + 2 \times 3 = 21$$

grandchildren. Thus, Mrs. Nicelady has $21 \times 2 = 42$ great grandchildren. Altogether, she has $5 + 21 + 42 = 68$ descendants.

Answer: 68

Problem 5 Solution

Note the signs of the numbers on the list alternate every other number, so the 12th number is positive. If we ignore the signs, we can see the numbers increase by 2 every time, so the 12th number is $11 \times 2 = 22$ more than 2, so 24.

Answer: 24

Problem 6 Solution

Since Derek had \$84 worth of gold when it was worth \$21 per ounce, he has $84 \div 21 = 4$ ounces of gold. This is $\frac{4}{12} = \frac{1}{3}$ pounds of gold. That same amount of gold is worth now $450 \times \frac{1}{3} = 150$ dollars.

Answer: 112.5

Problem 7 Solution

The difference in weight after they ate $\frac{2}{3}$ of the berries is

$$20 - 10 = 10$$

pounds. That means, $\frac{2}{3}$ of the berries weighed 10 pounds. Therefore, the berries alone weighed $10 \div \frac{2}{3} = 15$ pounds, and thus the backpack alone weighs $20 - 15 = 5$ pounds.

Answer: 5

Problem 8 Solution

The largest 3-digit number that can be formed with the given digits is 983 and the smallest 3-digit number that can be formed with the given digits is 238. Their difference is $983 - 238 = 745$.

Answer: 745

Problem 9 Solution

As the smallest two squares have perimeter 12 and 20, they have side length 3 and 5, respectively. Thus the second largest square has side length $3 + 5 = 8$, and the largest square has side length $5 + 8 = 13$. Therefore the largest square has perimeter $13 \times 4 = 52$.

Answer: 52

Problem 10 Solution

Consider the diagram below, where the whole rectangle is divided into 8 identical triangles.

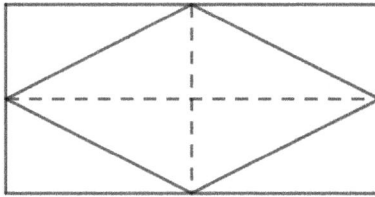

The rhombus is made up of 4 of these triangles, so each triangle has area $96 \div 4 = 24$.

Answer: 24

Problem 11 Solution

Since the number leaves no remainder when we divide it by 2 or 5, it must be a multiple of 10, so it ends in 0. The numbers bigger than 100 and smaller than 150 that end in 0 are 110, 120, 130 and 140. From these, the only one that leaves a remainder of 1 when divided by 7 is 120.

Answer: 120

Problem 12 Solution

Since the rectangle has perimeter 20 and one side of length 2, the other side has length

$$(20 - 2 - 2) \div 2 = 8.$$

Thus, the rectangle has area $2 \times 8 = 16$ and the square has area $4 \times 16 = 64$. As $8 \times 8 = 64$, the square has side length 8.

Answer: 8

Problem 13 Solution

If Norton answered all questions correctly, he would have a perfect score of $25 \times 6 = 150$ points.

For each incorrect or blank question he gets $6 + 1 = 7$ less points in his final score, since he loses the 6 points for having the correct answer and 1 point is deducted from his overall score.

The total decrease in his score was $150 - 101 = 49$, so he must have missed $49 \div 7 = 7$ questions.

Answer: 7

Problem 14 Solution

The M is made up of two rectangles with base 1 and height 5 and two parallelograms with base 1 and height 5. Note the volume of each of the 4 pieces is the same $1 \times 5 = 5$, so the whole figure has area $4 \times 5 = 20$.

Answer: 20

Problem 15 Solution

To buy $7 - 4 = 3$ cakes, Chris needs $12 + 21 = 33$ dollars. This means each cake costs $33 \div 3 = 11$ dollars.

Answer: 11

Problem 16 Solution

Since the bigger slices are twice as big as the smaller slices, we can think of them as 2 small slices of pie together. Thus, we can pretend Ethan sliced the pie into $18 + 2 = 20$ equal slices. There are 360 degrees in a full circle, so a slice that is $\dfrac{1}{20}$ of a circle would have an angle measure of $360 \div 20 = 18$ degrees.

Answer: 18

Problem 17 Solution

By the time the second train leaves the station, the first train already advanced $3 \times 70 = 210$ miles. Every extra hour, the second train travels $100 - 70 = 30$ more miles than the first train. So, to make up for the extra 210 miles the first train has traveled so far, they must travel for $210 \div 30 = 7$ more hours.

Answer: 7

Problem 18 Solution

The length of the small rectangles is equal to 5 times their width. This means the perimeter of one of the small rectangles is equal to $1 + 5 + 1 + 5 = 12$ times its width. Thus, the width of each small rectangle is $60 \div 12 = 5$ and their length is $5 \times 5 = 25$. The perimeter of the big rectangle is then

$$25 + 25 + 5 + 25 + 5 + 25 + 25 + 5 + 25 + 5 = 170.$$

Answer: 170

Problem 19 Solution

We know the days of the week repeat every 7 days. A common year has 365 days and a leap year has 366 days. The remainder of dividing 365 by 7 is 1, and the remainder of dividing 366 by 7 is 2.

This means after a common year passes, the same date the following year will occur 1 weekday later, and after a leap year passes, the same date the following year will occur 2 weekdays later. So, in the following years April 13th will be on

Year	2018	2019	2020	2021	2022	2023
April 13th	Fri	Sat	Mon	Tue	Wed	Thu
Year	2024	2025	2026	2027	2028	2029
April 13th	Sat	Sun	Mon	Tue	Thu	Fri

Therefore, next Friday April 13th will occur in $2029 - 2018 = 11$ years.

Answer: 11

Problem 20 Solution

The grape flavor tap can fill $\dfrac{1}{30}$ of the cup in one second, and the lime flavor tap can fill $\dfrac{1}{45}$ of the cup in one second, thus, if he opens both at the same time he can fill $\dfrac{1}{30} + \dfrac{1}{45} = \dfrac{5}{90} = \dfrac{1}{18}$ of the cup in one second. Therefore he would need to pour both flavors of soda for 18 seconds to fill his cup.

Answer: 18

2.7   ZIML April 2018 Division E

Below are the solutions from the Division E ZIML Competition held in April 2018.

The problems from the contest are available on p.63.

Problem 1 Solution

For every 1 blue flower Oliver collected 2 yellow flowers and 3 red flowers, that is, he had $1 + 2 + 3 = 6$ flowers in total for every 1 blue flower. Since he collected 78 flowers in total, he collected $78 \div 6 = 13$ blue flowers, and therefore collected $13 \times 3 = 39$ red flowers.

Answer: 39

Problem 2 Solution

There are 3 possible ways to attach one unit square to the figure so it has an axis of symmetry, and 3 possible ways to attach two unit squares to the figure so it has an axis of symmetry, as seen below.

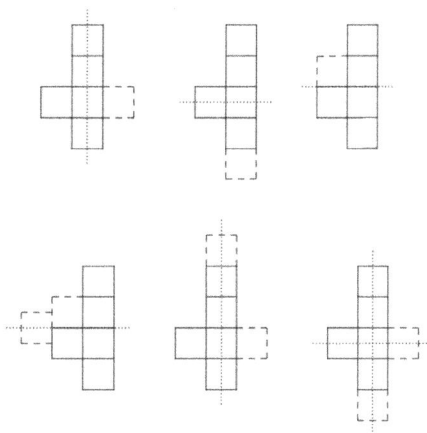

Answer: 6

Problem 3 Solution

Since we want Grassy to jump the least possible length, we want him to jump jumps of length 3 as much as possible. He does not jump the same length 3 times in a row, so he must jump 5 inches before/after he jumps 3 inches twice. Note $7 \div 3 = 2R1$, so he must jump 5 inches 2 times and 3 inches $2 \times 2 + 1 = 5$ times. This way, the shortest total length Grassy can jump is $5 \times 2 + 3 \times 5 = 25$ inches.

Answer: 25

Problem 4 Solution

Each week Liam usually eats $3 \times 7 = 21$ bananas. This week he ate $21 + 2 = 23$ bananas, so last week he ate $23 - 5 = 18$ bananas. Thus, during the past two weeks Liam ate $18 + 23 = 41$ bananas.

Answer: 41

Problem 5 Solution

Two faces of the triangular prism are right triangles with base 6 inches and height 8 inches, so they each have area $\dfrac{6 \times 8}{2} = 24$ square inches. The other three faces are rectangles that have area $15 \times 10 = 150$ square inches, $15 \times 6 = 90$ square inches, and $8 \times 15 = 120$ square inches. Thus, the surface area of the triangular prism is $24 + 24 + 150 + 90 + 120 = 408$ square inches.

Answer: 408

Problem 6 Solution

If we look at the differences between consecutive terms in the sequence,

$$2 - 3 = -1, 4 - 2 = 2, 1 - 4 = -3, 5 - 1 = 4,$$
$$0 - 5 = -5, 6 - 0 = 6, \ldots$$

we obtain the sequence

$$-1, 2, -3, 4, -5, 6, \ldots,$$

so the next number in the list should be 7 less than 6, that is $6 - 7 = -1$.

Answer: -1

Problem 7 Solution

If we rotate the small triangles to the right and left of the rectangle, we can come up with a rectangle that has the same area as the trapezoid and has two times the area of the shaded rectangle, as shown in the diagram.

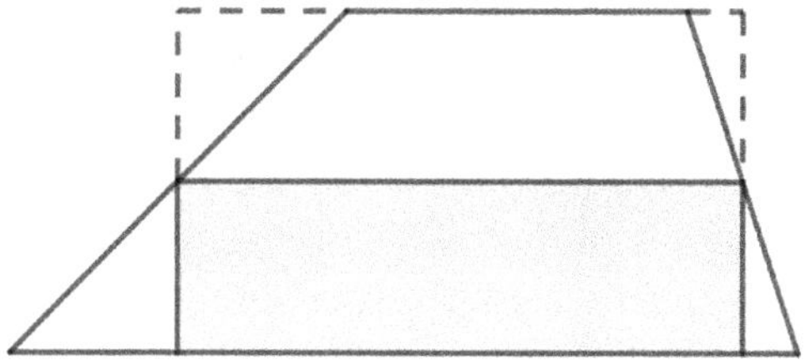

Thus the area of the trapezoid is $2 \times 23 = 46$.

Answer: 46

Problem 8 Solution

The area of a triangle with base b and height h is $A = \dfrac{b \times h}{2}$, so $13.5 \times 2 = 27$ is equal to the base times the height of her triangle. Note $27 = 3 \times 9$, so Jamie's triangle has base 9 and height 3.

Answer: 9

Problem 9 Solution

We work based on the hundreds digit.

If the hundreds digit of Mel's number was 0 (so the number is less than 100), the tens digit would be $3 \times 0 = 0$ and the ones digit would be $2 \times 0 = 0$, which cannot happen since Mel is thinking of a number between 1 and 1000.

If the hundreds digit was 1, the tens digit would be $3 \times 1 = 3$ and the ones digit would be $2 \times 1 = 2$, giving the number 132, which is $132 - 123 = 9$ more than when we swap its last two digits.

If the hundreds digit was 2, the tens digit would be $3 \times 2 = 6$ and the ones digit would be $2 \times 2 = 4$, giving the number 264, which is $264 - 246 = 18$ more than when we swap its last two digits.

If the hundreds digit was 3, the tens digit would be $3 \times 3 = 9$ and the ones digit would be $2 \times 3 = 6$, giving the number 396, which is $396 - 369 = 27$ more than when we swap its last two digits.

If the hundreds digit was 4 or more, the tens digit could not be 3 times that digit, so the only possible number is 264.

Answer: 264

Problem 10 Solution

If Ricky had answered all questions correctly, he would have had $5 + 5 = 10$ points, which is $10 - 6 = 4$ more points than he actually got. When Ricky answers correctly to a question, his score is $1 + 1 = 2$ more points than if he had answered incorrectly, which means he answered $4 \div 2 = 2$ questions incorrectly and $5 - 2 = 3$ questions correctly.

Answer: 3

Problem 11 Solution

Her basket weighs $21 - 9 = 12$ less pounds than it did before, so $\frac{2}{3}$ of her apples weigh 12 pounds. This means she originally had

$$12 \div \frac{2}{3} = 12 \times \frac{3}{2} = 18$$

pounds of apples in her basket.

Answer: 18

Problem 12 Solution

Looking at the first number, we see that 2 is changed to 5, 6 is changed to 7, etc. Combining all the information we get the table below:

Original	0	1	2	3	4	5	6	7	8	9
Substituted	4	3	5	9	1	6	7	0	8	2

(Note we had to use process of elimination to determine 4 is substituted by 1.) Therefore if Kiara received the number 82649, working backwards (substituted to original) we find the number Brendan actually sent was 89503.

Answer: 89503

Problem 13 Solution

All 4 triangles formed between the small square and the large square are congruent and have area $\frac{5 \times 15}{2} = \frac{75}{2}$. The side length of the bigger square is $15 + 5 = 20$, so it has area $20 \times 20 = 400$. The area of the smaller square is thus

$$400 - 4 \times \frac{75}{2} = 400 - 2 \times 75 = 250.$$

Answer: 250

Problem 14 Solution

The digit 5 appears on the ones place one time every 10 consecutive numbers from 155 to 365, so there are

$$36 - 15 + 1 = 22$$

of those. It also appears 10 times in the tens digit every 100 consecutive numbers, so there are $3 \times 10 = 30$ more 5's (150 to 159, 250 to 259, 350 to 359).

Hence in total Dustin wrote the digit 5 $22 + 30 = 52$ times.

Answer: 52

Problem 15 Solution

Between people spelling his name Giovanni and Jovan E., Jovani got his name spelled wrong $30 + 15 = 45$ percent of the time, that is, $60 \times 0.45 = 27$ times. Together with the times they spelled Joevany, they spelled his name wrong $27 + 15 = 42$ times. This means they spelled his name correctly $60 - 42 = 18$ times over the past three months.

Answer: 18

Problem 16 Solution

Since Natalie used 16 cups of diced apples, she baked

$$16 \div 2\frac{2}{3} = 16 \div \frac{8}{3} = 16 \times \frac{3}{8} = 6$$

pies. Hence she needed $1\frac{1}{2} \times 6 = 9$ cups of all-purpose flour.

Answer: 9

Problem 17 Solution

After drawing the blue, green and red lines, Patrick drawing looked like in the diagram below.

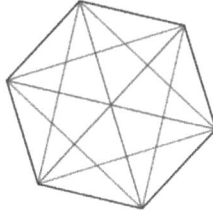

There are 2 big red triangles and 6 small red triangles, so there are $6 + 2 = 8$ red triangles in his drawing.

Answer: 8

Problem 18 Solution

Since 1.5 liters are equal to 1500 milliliters, Bailey can refill her glass of water $1500 \div 250 = 6$ times with a full pitcher. This means she can dance for $0.5 \times 6 + 0.5 = 3.5$ hours by the time she needs to refill her pitcher.

Answer: 3.5

Problem 19 Solution

Between 100 and 199 there are $199 - 100 + 1 = 100$ numbers, all with hundreds digit 1, so they are all divisible by their hundreds digit. Between 200 and 299 exactly half of the numbers are divisible by 2, so there are $100 \div 2 = 50$ such numbers. Since 300 is divisible by 3, there are in total $100 + 50 + 1 = 151$ numbers between 100 and 300 that are divisible by their hundreds digit.

Answer: 151

Problem 20 Solution

If it keeps raining at the same rate, after three hours the water will reach a height of 36 cm.

Thus, the volume of water in the bucket will be that of a rectangular prism of dimensions $30 \times 30 \times 18$, that is, 16200 cubic centimeters of water, which is the same as $16200 \div 1000 = 16.2$ liters of water.

Answer: 16.2

2.8 ZIML May 2018 Division E

Below are the solutions from the Division E ZIML Competition held in May 2018.
The problems from the contest are available on p.71.

Problem 1 Solution

Note that the signs of the numbers alternate every other number, and the value of the numbers (ignoring the minus sign) increases by 3 every time. So the 10th number on the sequence must be $3 \times 9 = 27$ more than 5, which is $5 + 27 = 32$.

Answer: 32

Problem 2 Solution

Recall a heptagon is a polygon with 7 sides. Each diagonal joins each pair on non-consecutive vertices, as shown on the diagram below.

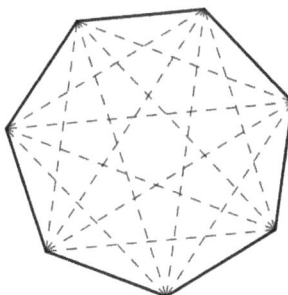

We can see there are $4 + 4 + 3 + 2 + 1 = 14$ diagonals in a regular heptagon.

Answer: 14

Problem 3 Solution

As he licks his lollipop 4 times every 7 seconds, Curt had to lick his lollipop a total of

$$637 \div 7 \times 4 = 364$$

times in total.

Answer: 364

Problem 4 Solution

We can break the perimeter of the figure into four pieces of length 5, one piece of length 3, one piece of length 4, and three pieces of length $4 - 3 = 1$. Thus, the perimeter of the figure is

$$4 \times 5 + 3 + 4 + 3 \times 1 = 30.$$

Answer: 30

Problem 5 Solution

Today there are 8 boys and 8 girls in the class. Since $4 + 3 = 7$ students were missing from class today, there are $8 + 8 + 7 = 23$ in Celeste's class.

Answer: 23

Problem 6 Solution

The perimeter of a rectangle with length l and width w is

$$P = 2 \times (l + w),$$

so the length and the width of her rectangle add to $16 \div 2 = 8$. Thus the length and width of her rectangle could be 1 and 7, 2 and 6, 3 and 5, or 4 and 4. So, Beth can draw 4 different rectangles.

Answer: 4

Problem 7 Solution

Since he will choose one of each kind of item, we can use the multiplication principle to count how many different ways he can choose. Thus he has $5 \times 3 \times 5 \times 3 = 225$ different ways to choose his outfit.

Answer: 225

Problem 8 Solution

25 years ago the sum of the ages of Dan and Han was

$$56 - 25 - 25 = 6,$$

so Han was $6 \div (1 + 2) = 2$ years old and Dan was $2 \times 2 = 4$ years old. This means today Han is $2 + 25 = 27$ years old (and Dan is $4 + 25 = 29$ years old).

Answer: 27

Problem 9 Solution

Without counting the door and the window, they need to cover 2 walls of area $9 \times 8 = 72$ square feet, and another 2 walls of area $12 \times 8 = 96$ square feet.

The area of the door is $1.5 \times 7 = 10.5$ square feet, and the area of the window is $3 \times 6 = 18$ square feet.

Thus, they will need

$$2 \times 72 + 2 \times 96 - 10.5 - 18 = 307.5$$

square feet of wallpaper.

Answer: 307.5

Problem 10 Solution

Since the same kid counted the number 2 and 44, that means the last kid on the circle counted the number $44 - 2 = 42$. So, the number of kids sitting on the grass must be a factor of 42. The only odd factors of 42 are 1, 3, 7, and 21, so there were 7 kids sitting on the grass.

Answer: 7

Problem 11 Solution
Since he jumped 0.5 meters in his last jump, the previous three jumps were 1 meter long, 2 meters long, and 4 meters long. Thus, he jumped $4 + 2 + 1 + 0.5 = 7.5$ meters in total.

Answer: 7.5

Problem 12 Solution
Since he eats the same number of cookies each of the 7 days of the week, the number of cookies he gets every Sunday is 4 more than a multiple of 7, 11 more than a multiple of 7, and 18 more than a multiple of 7.

Noticing that $4 + 7 = 11$ and $11 + 7 = 18$ we see that it works if Duncan gets $18 + 7 = 25$ cookies each week. As Duncan eats at least one cookie each day, this is the minimum number possible.

Answer: 25

Problem 13 Solution
Corrine has a total of $20 \times 2 = 40$ slices of bread and $4 \times 12 = 48$ slices of smoked turkey.

She needs two slices of bread for each sandwich, so she has enough bread for $40 \div 2 = 20$ sandwiches, and she has enough slices of smoked turkey for $48 \div 3 = 16$ sandwiches.

Thus, she can make at most 16 sandwiches.

Answer: 16

Problem 14 Solution

To fill the pool he would need $8 \times 20 \times 7 = 1120$ cubic feet of water. That is about $1120 \div 10 \times 75 = 8400$ gallons of water.

Answer: 8400

Problem 15 Solution

The hundreds digit of the number of candies he gets will be half of 6, so 3. The tens digit will be 5. Lastly, the ones digit will be two times 4, so 8.

Thus, he will get 358 candies.

Answer: 358

Problem 16 Solution

The base and the height of the triangles add up to the side length of the square. Since the base of the triangle is half the height, the base is
$$12 \div (1 + 2) = 4$$
and the height is $2 \times 4 = 8$.

Thus, the area of each of the triangles is $4 \times 8 \div 2 = 16$, and the shaded area is $3 \times 16 = 48$.

Answer: 48

Problem 17 Solution

The only numbers that divide 33 are 1, 3, 11 and 33, so one of those must be the number of hens in Lindy's farm.

There can't be 33 hens, as there are 20 hens and roosters in total, and if there were 1 or 3 hens, there would be more roosters than hens, so there are 11 hens and hence $20 - 11 = 9$ roosters in Lindy's farm.

Answer: 9

Problem 18 Solution

We can cut and rearrange the shaded area like in the diagram below.

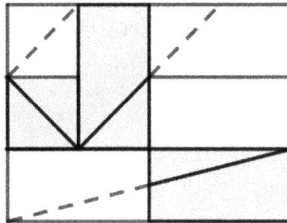

Rearranged, the shaded area covers two and a half of the rectangles that form the figure. Thus, the shaded area is

$$2.5 \div 5 = 0.5 = 50\%$$

of the figure.

Answer: 50

Problem 19 Solution

At first Grace ate

$$50 \times 20\% = 50 \times 0.2 = 10$$

cookies, so $50 - 10 = 40$ cookies were left in the jar.

Then she ate
$$40 \times 20\% = 40 \times 0.2 = 8$$
cookies. Thus Grace ate a total of $10 + 8 = 18$ cookies.

Answer: 18

Problem 20 Solution

The number must be 7 more than a multiple of 12.

The largest 2-digit multiple of 12 is 96, but 7 more than that is $96 + 7 = 103$, which is not a 2-digit number.

The next largest 2-digit number of 12 is 84, thus the number we are looking for is $84 + 7 = 91$.

Answer: 91

2.9   ZIML June 2018 Division E

Below are the solutions from the Division E ZIML Competition held in June 2018.

The problems from the contest are available on p.77.

Problem 1 Solution

There are 12 cubes touching the wall on the left; next to those there are 6 cubes; then 2 more cubes; and 4 more in the L shape. Thus, there are in total $12 + 6 + 2 + 4 = 24$ cubes.

Answer: 24

Problem 2 Solution

Since Addie used 905 instead of 506, the answer she got was $905 - 506 = 399$ more than the correct one.

Thus, the correct answer is

$$65536 - 399 = 65536 - 400 + 1 = 65137.$$

Answer: 65137

Problem 3 Solution

For each of her kids Dustina will buy

$$1 + 2 + 1 + 1 + 1 + 1 + 10 + 3 + 2 + 1 + 1 = 24$$

items. Since she is buying 11 different items, she will get in total $2 \times 24 + 11 = 59$ items.

Answer: 59

Problem 4 Solution

If we look at the "flower" figure formed in the middle of the figure, there are 4 lines of symmetry by looking in between the petals, and 4 more by slicing the "petals" in the middle, as shown in the diagram below.

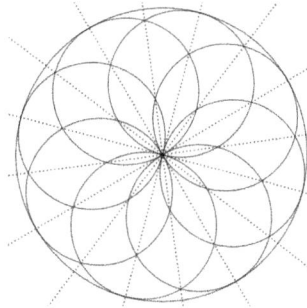

Thus, there are $4 + 4 = 8$ lines of symmetry in the figure.

Answer: 8

Problem 5 Solution

Daria needs to be extremely careful with the order of operations! She will be able to find the biggest number by first multiplying the two biggest numbers and then subtracting the smallest one. So, she should do $8 \times 5 - 2 = 38$ or $5 \times 8 - 2 = 38$.

Answer: 38

Problem 6 Solution

If all of the 45 items in storage were bicycles, there would have been $2 \times 45 = 90$ wheels in total. Each tricycle brings $3 - 2 = 1$ extra wheel, which means each of the extra $118 - 90 = 28$ wheels actually corresponds to a tricycle. Therefore, there were 28 tricycle in storage.

Answer: 28

Problem 7 Solution

A number that leaves a remainder of 2 when divided by 5 always ends in either 2 or 7, so the 3-digit numbers are $102, 107, 112, \ldots$.

Note that 105 is a multiple of 3, so the first number in our list that has a remainder of 2 when divided by 3 is 107. Thus 107 is the number we are looking for.

Answer: 107

Problem 8 Solution

If Dan had made 4 more floral arrangements, they would have made $44 + 4 = 48$ floral arrangements in total, and each would have made the same number.

This means, Sam made $48 \div 2 = 24$ floral arrangements in $6 \times 60 = 360$ minutes. So, on average, he spent $360 \div 24 = 15$ minutes making each floral arrangement.

Answer: 15

Problem 9 Solution

Since the rectangle is three times as long as it is wide, we can split it into three smaller squares, each with area $75 \div 3 = 25$.

Thus, the width of the rectangle is 5 (since $5 \times 5 = 25$ is the area of each square), and the length of the rectangle is $3 \times 5 = 15$.

Therefore, Don needs $5 + 15 + 5 + 15 = 40$ feet of fence.

Answer: 40

Problem 10 Solution

If we add together all the weights we know, we will get the total weight of 2 boxes of each kind of nail. Thus, 2 boxes of each kind of nail together weigh $8 + 18 + 14 = 40$ pounds. This means, together, 1 box of each kind of nail weighs $40 \div 2 = 20$ pounds. As one box of super small nails and one box of big nails weigh 14 pounds together, one box of small nails weighs $20 - 14 = 6$ pounds.

Answer: 6

Problem 11 Solution

Note the length of the line is equal to the sum of the side lengths of all squares. The perimeter of each square is 4 times its side length, so the sum of the perimeters of all squares is equal to 4 times the length of the line. Therefore, the sum of the perimeter of all squares is $13 \times 4 = 52$.

Answer: 52

Problem 12 Solution

Since Isabella filled 43 lines with 91 characters each, she wrote $43 \times 91 = 3913$ characters. Knowing the word Hippopotomonstrosesquipedaliophobia has 35 characters, and $3913 \div 35 = 111$ with remainder 28, she wrote the word 111 times and then the first 28 characters of the word. As the first 28 characters contain the letter p four times, she wrote the letter p a total of $5 \times 111 + 4 = 559$ times.

Answer: 559

Problem 13 Solution

The rectangles with area 6 and 15 share one side, and the only common factors of 6 and 15 are 1 and 3. The rectangles with area 15 and 50 share one side, and the only common factors of 15 and 50 are 1, and 5.

Thus, the rectangle with side area 15 has dimensions 5×3, the one with area 6 has dimensions 2×3, and the one with area 50 has dimensions 5×10. Thus the fourth rectangle has dimensions 2×10, and so it has area 20. Therefore, the area of the entire rectangle is $6 + 15 + 50 + 20 = 91$.

Answer: 91

Problem 14 Solution

Each of the bottles will hold $3.78 \div 2 = 1.89$ liters of milk. Karina has available $15 \times 5 = 75$ liters of milk, so she will be able to fill

$$75 \div 1.89 \approx 39.68$$

bottles of milk. Therefore Karina can fill 39 bottles of milk completely.

Answer: 39

Problem 15 Solution

We can make a list of all the ways of ordering the digits 1 through 4 in a 4 digit number making sure that, it does not start with 1, 2 is not on the second place, 3 is not on the third place and 4 is not on the last place, while making sure to use all 4 numbers. This way we can represent all possible combinations so that no one buys a gift form themselves. (A nice way of coming up with this list is to make up a tree diagram). The list of numbers we want is

$$
\begin{array}{ccc}
2143 & 2341 & 2413 \\
3142 & 3412 & 3421 \\
4123 & 4312 & 4321
\end{array}
$$

so there are 9 different ways in which they can give presents to each other.

Answer: 9

Problem 16 Solution

Note a pattern in the number of matches used: A triangle of side length 1 uses 3 matchsticks; a triangle of side length 2 uses 9 matchsticks; a triangle with side length 3 uses 18 matchsticks.

Each time the total number of matchsticks increases by as many matchsticks as the perimeter of a triangle of that size: $3, 3+6 = 9$, $3+6+9 = 18, \ldots$. Thus, a triangle with side length 7 will need

$$3+6+9+12+15+18+21 = 84$$

matchsticks.

Answer: 84

Problem 17 Solution

Recall 100 pennies, 4 quarters, 10 times, 20 nickels, and 1 silver dollar, are all worth exactly $1. So, Inez had in total

$$\begin{aligned}
&358 \div 100 + 172 \div 4 + 236 \div 10 + 50 \div 20 + 15 \div 1 \\
&= 3.58 + 43.00 + 23.60 + 2.50 + 15 \\
&= 87.68
\end{aligned}$$

dollars. So, she spent $87.68 - 33.68 = 58$ dollars for the 3 movie tickets. Thus, each movie ticket cost $54 \div 3 = 18$ dollars.

Answer: 18

Problem 18 Solution

The first 10 even numbers are 2, 4, ..., 20. Which are the same as 2 times each of 1, 2, ..., 10. The largest powers of prime factors that are factors of one or more of the first 10 integers are $2^3 = 8$,

$3^2 = 9, 5^1 = 5$, and $7^1 = 7$. Hence, the least common multiple of the first 10 even numbers is $2 \times 2^3 \times 3^2 \times 5 \times 7 = 5040$.

Answer: 5040

Problem 19 Solution

60% of the respondents were female, that is, $200 \times 60\% = 120$ respondents. Out of these 120 female respondents $70 - 14 = 56$ answered "Don't know". So $120 - 40 - 56 = 24$ female respondents answered "No" to the question.

Therefore, $24 \div 120 = 0.2 = 20\%$ of the female respondents would not drink the rain if it was made out of chocolate.

Answer: 20

Problem 20 Solution

The different ways to make a square using 1, 2, 3 or 4 of the pieces are shown below.

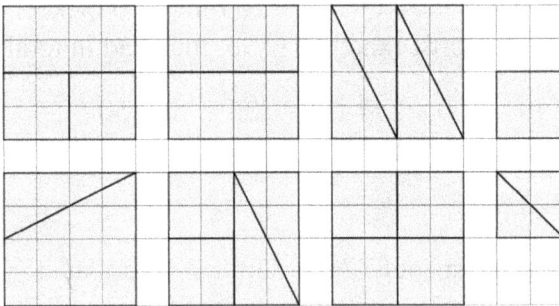

Thus, there are 8 different ways of making a square using the given figures.

Answer: 8

3. Appendix

3.1 Division E Topics Covered

Note: Setting up and solving equations is not necessary for the problems in Division E (there are rare exceptions among the earlier monthly contests though). Students are allowed to use equations to solve the questions, but the questions are designed to be solved without using equations or systems of equations.

Word Problems

- Calculations and Arithmetic: Adding, subtracting, multiplying, and dividing whole numbers, fractions, and decimals
- Ratios and Proportions: Using ratios to find parts of a whole, Calculating missing information from proportional relationships, etc.
- Percents: Calculating percent increases and decreases, Relationship between percents and ratios, Using percents in mixture problems (e.g. 40% water and 60% oil)
- Problem Solving Methods: Chicken and Rabbit method, Using ratios when given sums or differences
- Motion Problems using (Speed)x(Time)=(Distance), Average Speed, Applying proportions to motion problems

---

- Work using (Rate)x(Time)=(Work Done), Average Rate of Work, Applying proportions to work problems

Geometry

- Areas and Perimeters of Basic Shapes such as triangles, rectangles, parallelograms, trapezoids, and circles
- Symmetry of Polygons
- Similar Triangles: Equal Angles, Sides are in a common ratio
- Geometric Reasoning with Areas: Congruent shapes have the same area, Dividing a shape and rearranging areas to find patterns, etc.
- Volumes and Surface Areas of Basic Solids such as cubes and rectangular prisms (boxes)

Number Sense

- Place Values: Ones/units digit, tens digit, hundreds digit, etc.
- Fundamental Definitions: Quotients and Remainders, Prime numbers, Factors (Divisors), Multiples, Perfect squares, Perfect cubes, etc.
- Least common multiple (LCM), Greatest common factor or divisor (GCF or GCD)
- Sum and Product Rules for Counting
- Sequences: Arithmetic and Geometric Sequences, Sum of elements in an arithmetic sequence, Finding patterns for general sequences
- Probability: Gives the chance of something happening, Ratio of outcomes
- Basic Statistics: Mean (Average), Median, Mode for lists, Interpreting data from graphs, bar charts, tables, etc.

3.2 Glossary of Common Math Terms

Acute Angle An angle less than $90°$.

Altitude of a Triangle A line segment connecting a vertex of a triangle to the opposite side forming a right angle. Also called the height of a triangle.

Angle A figure formed by two rays sharing a common vertex. Often measured in degrees.

Arc The curve of a circle connecting two points.

Area The amount of space a region takes up. Often denoted using square brackets: area of $\triangle ABC = [ABC]$.

Arithmetic Sequence A sequence where the difference between one term and the next is constant.

Average See Mean.

Base of a Triangle One side of a triangle, often used when the altitude is drawn from the opposite side to this base.

Chord A line segment connecting two points on the outside of a circle.

Circle A round shape consisting of points that all have the same distance (called the radius) from the center of the circle.

Circumference The perimeter of a circle.

Composite Number A number that is not prime.

Congruent Two shapes or figures that are exactly the same.

Cube A solid figure formed by 6 congruent squares that all meet at right angles.

Deck of Cards A standard deck of cards has 52 cards. There are 4 suits (clubs, diamonds, hearts, and spades) with each suit having cards of 13 ranks (A (ace), $2, 3, \ldots, 10$, J (jack), Q (queen), and K (king)).

Denominator The bottom number in a fraction.

Diagonal A line segment connecting two vertices of a shape or solid that is not an edge of the shape or solid.

Diameter A chord passing through the center of a circle. The diameter has length that is twice the radius.

Die or Dice A standard die (plural is dice) has 6 sides. Each of the 6 sides has the same chance when the die is rolled.

Digit One of $0, 1, 2, \ldots, 9$ used when writing a number.

Distinguishable Objects Objects that are different.

Divisible A number is divisible by another number if there is no remainder when the first number is divided by the second. For example, 35 is divisible by 7.

Divisor A number that evenly divides another number. For example, 6 is a divisor of 48. Also called a factor.

Edge A line segment connecting two vertices on the outside of a shape or solid.

Equally Likely Having the same chance of occurring.

Equiangular Polygon A shape with all equal angles.

Equilateral Polygon A shape with all equal sides.

Equilateral Triangle A regular triangle, one with three equal sides and three equal angles.

Even Number A number divisible by 2.

Exponent The number another number is raised to for powers. For example, in a to the power of b (a^b), the exponent is b.

Face The shape or polygon on the outside of a solid region.

Factor of a Number A number that evenly divides another number. For example, 6 is a factor of 48. Also called a divisor.

Factorial The symbol ! where $n! = n \times (n-1) \times (n-2) \cdots \times 1$.

Fraction An expression of a quotient. For example, $\dfrac{1}{2}$ or $\dfrac{9}{7}$.

Geometric Sequence A sequence where the ratio between one term and the next is constant.

Greatest Common Divisor (GCD) The largest number that is a divisor/factor of two or more numbers.

Greatest Common Factor (GCF) See Greatest Common Divisor.

Indistinguishable Objects Objects that are the same.

Isosceles Triangle A triangle with two equal sides and two equal angles.

Least Common Multiple (LCM) The smallest number that is a multiple of two or more numbers.

Mean The sum of the numbers in a list divided by the how many numbers occur in the list. Also called the average.

Median The number in the middle of a list when the list is arranged in increasing order.

Midpoint The point in the middle of a line segment.

Mode The number or numbers occurring most often in a list of numbers.

Multiple A number that is an integer times another number. For example, 72 is a multiple of 8.

Numerator The top number in a fraction.

Obtuse Angle An angle between $90°$ and $180°$.

Odd Number A number not divisible by 2.

Parallel Lines Lines that do not intersect.

Perfect Cube A number that is another number cubed. For example, $64 = 4^3$ is a perfect cube.

Perfect Square A number that is another number squared. For example, $64 = 8^2$ is a perfect square.

Perimeter The length/distance around the outside of a shape.

Pi (π) A number used often in geometry. $\pi = 3.1415926\ldots \approx 3.14 \approx \dfrac{22}{7}$.

Polygon A shape formed by connected line segments.

Prime Factorization The expression of a number as the product of all its prime factors. For example, 24 has prime factorization $2 \times 2 \times 2 \times 3 = 2^3 \times 3$.

Prime Number A number whose only factors are one and itself.

Proportional Ratios Ratios that have equal values when expressed in fraction form. For example, $2 : 3$ is proportional to $8 : 12$.

Quadrilateral A shape with four sides.

Quotient The integer quantity when dividing one number by another. For example, the quotient of $38 \div 5$ is 7 as $38 = 7 \times 5 + 3$.

Radius of a Circle The distance from the center of the circle to any point on the outside of the circle.

Randomly Chosen for a group of objects. Unless specified, the chance of choosing each object is the same as any other object.

Rank of a Card See Deck of Cards.

Ratio A relation depicting the relation between two quantities. For example $2 : 3$ or $\frac{2}{3}$ denotes that for every 3 of the second quantity there are 2 of the first quantity.

Rectangle A quadrilateral with four right angles (an equiangular quadrilateral).

Regular Polygon A polygon with all equal sides and all equal angles (equilateral and equiangular).

Remainder The quantity left over when one integer is divided by another. For example, the remainder of $38 \div 5$ is 3 as $38 = 7 \times 5 + 3$.

Rhombus A quadrilateral with four equal sides (an equilateral quadrilateral).

Right Angle A $90°$ angle.

Right Triangle A triangle containing a right angle.

Scalene Triangle A triangle with three unequal sides and three unequal angles.

Sequence An ordered list of numbers.

Similar Shapes or solids that have the same angles and sides that share a common ratio.

Square A shape with four equal sides and four equal angles (a regular quadrilateral).

Suit of a Card See Deck of Cards.

Surface Area The total area of all the faces of a solid.

Trapezoid A quadrilateral with one pair of parallel sides.

Triangle A shape with three sides.

Vertex The intersection of line segments, especially the intersection of sides or edges in a shape or solid.

Volume The amount of space a solid region takes up.

With Replacement When choosing objects with replacement, a chosen object is returned to the others allowing it to be chosen more than once.

3.3 ZIML Answers

ZIML October 2017 Division E

Problem 1:	10	Problem 11:	720
Problem 2:	13	Problem 12:	192
Problem 3:	7	Problem 13:	15
Problem 4:	90	Problem 14:	300
Problem 5:	8	Problem 15:	89
Problem 6:	40	Problem 16:	6200
Problem 7:	6	Problem 17:	128
Problem 8:	444	Problem 18:	11
Problem 9:	114	Problem 19:	105
Problem 10:	30	Problem 20:	6

ZIML November 2017 Division E

Problem 1:   34   Problem 11:   22

Problem 2:   54   Problem 12:   12

Problem 3:   245   Problem 13:   80

Problem 4:   240   Problem 14:   60

Problem 5:   86   Problem 15:   6

Problem 6:   3   Problem 16:   1

Problem 7:   20   Problem 17:   105

Problem 8:   15   Problem 18:   5535

Problem 9:   10   Problem 19:   15750

Problem 10:   40   Problem 20:   12

ZIML December 2017 Division E

Problem 1:	728	Problem 11:	10
Problem 2:	824	Problem 12:	40
Problem 3:	50	Problem 13:	64
Problem 4:	6	Problem 14:	8
Problem 5:	100	Problem 15:	3
Problem 6:	5	Problem 16:	17
Problem 7:	15	Problem 17:	4
Problem 8:	100	Problem 18:	40
Problem 9:	108	Problem 19:	36
Problem 10:	9	Problem 20:	9

ZIML January 2018 Division E

Problem 1:	40	Problem 11:	400
Problem 2:	100	Problem 12:	223
Problem 3:	8	Problem 13:	18
Problem 4:	32	Problem 14:	6
Problem 5:	82	Problem 15:	92
Problem 6:	23	Problem 16:	30
Problem 7:	20	Problem 17:	46
Problem 8:	250	Problem 18:	20
Problem 9:	6	Problem 19:	12
Problem 10:	187	Problem 20:	10

ZIML February 2018 Division E

Problem 1:	20376	Problem 11:	72
Problem 2:	9	Problem 12:	86
Problem 3:	12	Problem 13:	8
Problem 4:	5	Problem 14:	70
Problem 5:	6	Problem 15:	54
Problem 6:	15	Problem 16:	18
Problem 7:	6	Problem 17:	450
Problem 8:	12	Problem 18:	1890
Problem 9:	8	Problem 19:	121
Problem 10:	3	Problem 20:	10

ZIML March 2018 Division E

Problem 1: 112

Problem 2: 4

Problem 3: 605

Problem 4: 68

Problem 5: 24

Problem 6: 112.5

Problem 7: 5

Problem 8: 745

Problem 9: 52

Problem 10: 24

Problem 11: 120

Problem 12: 8

Problem 13: 7

Problem 14: 20

Problem 15: 11

Problem 16: 18

Problem 17: 7

Problem 18: 170

Problem 19: 11

Problem 20: 18

ZIML April 2018 Division E

Problem 1:	39	Problem 11:	18
Problem 2:	6	Problem 12:	89503
Problem 3:	25	Problem 13:	250
Problem 4:	41	Problem 14:	52
Problem 5:	408	Problem 15:	18
Problem 6:	-1	Problem 16:	9
Problem 7:	46	Problem 17:	8
Problem 8:	9	Problem 18:	3.5
Problem 9:	264	Problem 19:	151
Problem 10:	3	Problem 20:	16.2

ZIML May 2018 Division E

Problem 1:	32	Problem 11:	7.5
Problem 2:	14	Problem 12:	25
Problem 3:	364	Problem 13:	16
Problem 4:	30	Problem 14:	8400
Problem 5:	23	Problem 15:	358
Problem 6:	4	Problem 16:	48
Problem 7:	225	Problem 17:	9
Problem 8:	27	Problem 18:	50
Problem 9:	307.5	Problem 19:	18
Problem 10:	7	Problem 20:	91

ZIML June 2018 Division E

Problem 1:	24	Problem 11:	52
Problem 2:	65137	Problem 12:	559
Problem 3:	59	Problem 13:	91
Problem 4:	8	Problem 14:	39
Problem 5:	38	Problem 15:	9
Problem 6:	28	Problem 16:	84
Problem 7:	107	Problem 17:	18
Problem 8:	15	Problem 18:	5040
Problem 9:	40	Problem 19:	20
Problem 10:	6	Problem 20:	8

www.ingramcontent.com/pod-product-compliance
Lightning Source LLC
Chambersburg PA
CBHW050124210326
41519CB00015BA/4101